Using the Master's Tools

Using The Master's Tools

Resistance and the Literature of the African and South-Asian Diasporas

Anuradha Dingwaney Needham

St. Martin's Press
New York

ISBN 0-312-22542-3

Library of Congress Cataloging-in-Publishing Data
Needham, Anuradha Dingwaney, 1951-
Using the master's tools : resistance and the literature of the African and South-Asian
diasporas / by Anuradha Dingwaney Needham.
p. cm.
Includes bibliographical references and index.
ISBN 0-312-22542-3
1. English literature—Minority authors—History and criticism. 2. Literature and society—English-speaking countries—History—20th century. 3. James, C. L. R. (Cyril Lionel Robert), 1901—Criticism and interpretation. 4. English literature—20th century—History and criticism. 5. Aidoo, Ama Ata, 1942—Criticism and interpretation. 6. Rushdie, Salman—Criticism and interpretation. 7. Kureishi, Hanif—Criticism and interpretation. 8. Cliff, Michelle—Criticism and interpretation. 9. Postcolonialism—Great Britain—History—20th century. 10. Decolonization in literature. 11. Ethnic groups in literature. 12. Imperialism in literature. 13. Minorities in literature. 14. Colonies in literature. 15. Race in literature. I. Title.
PR120.M55 N44 2000
820.9'920693—dc21 99–054551

Design by Letra Libre, Inc.

First edition: May, 2000
10 9 8 7 6 5 4 3 2 1

For
Larry, Nalin, and Jayesh

Contents

Acknowledgments

During the nearly six years in which this book has been in the making, as it were, I've accumulated a number of debts—personal, institutional, professional—that I now wish to acknowledge. Katherine Linehan and Lawrence Needham read the entire manuscript and offered many constructive suggestions and a great deal of encouragement for which I'm enormously grateful. Paula Richman started me off on this project and was there toward the end, encouraging me to complete this book *before* I devoted myself to other projects. Rajeswari Sunder Rajan read and commented extensively on the introduction and the last chapter. Wendy Motooka read and commented on the introduction. Khachig Tölölyan's suggestions on chapter one, a version of which he first published in *Diaspora,* were indispensable, as was Agha Shahid Ali's help. Michael Fisher read and commented extensively on chapter two. Conversations with bell hooks, Mary Layoun, Áine O'Brien, Indira Karamcheti, Ama Ata Aidoo, and R. Radhakrishnan on different aspects of this book while it was being written were a continuous source of intellectual stimulation and insight. To all of them, a big thank-you.

This book quite literally grew out of my interactions with students and theirs with me, both in and outside the classroom. Without their astute questioning, it's fair to say, this would have been a different, much impoverished book. Special thanks go to my research assistants—Kai Lindstrom Williams, Doug Choi, Andy Comer; the "London" group on whom I tried so many of the ideas in this book—Mish Kim, Nina Spensley, Ji Um, Sabeena Sethia, Rachel Schiller, Devan Apthekar, Amy Pak; and finally those who did honors projects with me and from whom I learned so much—Rebekah Rutkoff, Jeanne Park, Matt Covey, Narges Kakalia, and Vince Schleitweiler.

The seed for this book was planted in a seminar I took with Edward W. Said at the School for Criticism and Theory at Dartmouth College. It took a while for me to get going, but Said's work has been a constant source of inspiration. Gayatri Chakravorty Spivak's strictures about the uses and limits of many of the positions I elaborate were particularly important. And then there is the enormously productive work on resistance by scholars too numerous to mention but whose impact on this work is clearly discernible in the epigraphs, notes, and citations. I could not have written this book without their enabling work on the subject.

A grant for College Teachers from the National Endowment for Humanities in 1992–1993 gave me time off to begin work on this book, and a Research Status fellowship from Oberlin College in 1997–1998 helped me complete it. I gratefully acknowledge this support. Along the way, help in the form of research assistants was extended by the Deans of the College of Liberal Arts and Sciences, Oberlin College—Alfred Mackay at the beginning and Clayton Koppes as the project neared its end—for which I'm very grateful. Clayton Koppes also provided funds for the index.

Eva Owens and Nalin Needham worked long hours compiling the index in record time, for which they deserve my heartfelt thanks.

My thanks also to all those in the reference section in Mudd Library—Helen Black, Jessica Grim, and Haiping Li, in particular—for hunting down references and meeting my requests for interlibrary loans with efficiency and speed. Eric Carpenter, too, deserves special mention for getting hold of a book I desperately needed and processing the paperwork in record time.

Versions of chapters 1 and 2 appeared first in *Diaspora: A Journal of Transnational Studies* and *The Yearly Review* (University of Delhi) and are reprinted with their permission.

For someone like me who has no clue about how a computer works, Barbara Finnegan and Linda Pardee, administrative assistants in the English Department, were indispensable. My thanks also to David Walker, chair of the English Department, for willingly allotting me departmental resources to draw upon at a time it mattered.

The people at St. Martin's Press—Kristi Long and Amy Reading—were generous in their support and time, for which I am very thankful. Kristi provided one of the most thorough critical readings my manuscript received, and her suggestions, always specific, were right on the mark. Amy metaphorically held my hand and calmed me down even as she responded to all my crazy queries. The anonymous reviewer helped enormously with comments that helped me in articulating clearly and well the rationale for this book; for this I am grateful.

Without Larry, my most generous and stringent reader, this work wouldn't have been done; and without Nalin and Jayesh, my sons, whose absolute impatience with anything that interfered with their schedule spurred me on to finish, I would most likely have still been working on this book. To them, in gratitude, I dedicate this book.

Permissions

ANTHONY BARNETT Excerpt from "After Nationalism" by Anthony Barnett, in *Patriotism: The Making and Un-Making of British National Identity, Volume I: History and Politics,* pp. 140–155. Edited by Raphael Samuel. Published by Routledge, New York, 1989; reprinted by permission of Taylor & Francis, for non-exclusive world rights in the English language.

MICHELLE CLIFF Excerpts from *The Land of Look Behind: Prose and Poetry* © 1985 by Michelle Cliff. Reprinted with the permission of Firebrand Books, Ithaca, New York, for non-exclusive world rights in the English language.

JONATHAN DOLLIMORE Excerpt from Jonathan Dollimore's *Sexual Dissidence: Augustine to Wilde, Freud to Foucault* © 1991, is reprinted by permission of Oxford University Press, for non-exclusive world rights in the English language.

FRANTZ FANON Excerpts from "The Pitfalls of National Consciousness" in *The Wretched of the Earth* ©1966, are reprinted with the permission of Grove/Atlantic Press, Inc. for non-exclusive world rights in the English language.

SIMON GIKANDI Quotation from *Maps of Englishness: Writing Identity in the Culture of Colonialism* by Simon Gikandi. Copyright © 1996 Columbia University Press. Reprinted with the permission of the publisher for non-exclusive world rights in the English language.

DICK HEBDIGE Excerpt from "Digging for Britain: an Excavation in Seven Parts" by Dick Hebdige, in *Come on Down: Popular Media Culture in Postwar Britain.* Edited by Dominic Strinatti and Stephen Wagg. Published by Routledge, London, 1992; reprinted by permission of Taylor & Francis, for non-exclusive world rights in the English language.

BELL HOOKS Quotation from "Narratives of Struggle." Reprinted by permission of the author of this article.

C. L. R. JAMES Excerpt from *Beyond a Boundary* by C. L. R. James. Published by Stanley Paul, London, 1963. Reprinted by permission of the Random House Archive & Library, a division of the Random House Group Ltd., for non-exclusive world rights in the English language.

HANIF KUREISHI Excerpts from *The Buddha of Suburbia* by Hanif Kureishi. Copyright © 1990 by Hanif Kureishi. Used by permission of Penguin, a division of Penguin Putnam, Inc., for rights in the United States and Canada. Reprinted with the permission of Faber and Faber Ltd., for non-exclusive rights throughout the United Kingdom and its domain in the English language.

Excerpts from *London Kills Me* by Hanif Kureishi. Copyright © 1992 by Hanif Kureishi. Used by permission of Viking Penguin, a division of Penguin Putnam, Inc., for rights in the United States and Canada. Reprinted with the permission of Faber and Faber Ltd., for non-exclusive rights throughout the United Kingdom and its domain in the English language.

ANURADHA DINGWANEY NEEDHAM Article entitled "Inhabiting the Metropole: C. L. R. James and the Postcolonial Intellectual of the African Diaspora" *Diaspora* 2:3 (1993). Reprinted by permission of University of Toronto Press Incorporated, for non-exclusive world rights in the English language.

EDWARD SAID Excerpt from "Intellectuals in the Postcolonial World" by Edward Said, in *Salmagundi* 70/71 (1986), pp. 44–64. Reprinted by permission of the author of this article.

GAYATRI CHAKRAVORTY SPIVAK Copyright © 1993. Excerpts from *Outside in the Teaching Machine* by Gayatri Chakravorty Spivak. Reproduced by permission of Taylor & Francis/Routledge, Inc., for non-exclusive world rights in the English language.

PAUL WILLEMEN Quotation from *Looks and Frictions: Essays in Cultural Studies and Film Theory* © 1994 is reprinted with the permission of Indiana University Press for exclusive rights in the United States, Puerto Rico, the Philippines, Canada, Mexico, and Japan in the English language. Reprinted with the permission of the British Film Institute for exclusive rights in the United Kingdom and its domain in the English language.

SYLVIA WYNTER Quotation from "The Ceremony Must Be Found: After Humanism," *boundary 2,* 12:3/13:1, Spring/Fall 1984) © 1984 is reprinted by permission of Duke University Press for non-exclusive world rights in the English language. All rights reserved.

Introduction

> *In this book [*Outside in the Teaching Machine*] I have repeatedly emphasized the formula of the new politics of reading: not to excuse, not to accuse, establish critical intimacy, use (or ab-use) the seeming weak moments for scrupulous ends.*
>
> —Gayatri Chakravorty Spivak

> *[Walter] Benjamin was always reluctant to assume individual phenomena under general concepts and advocated that the writer [and also the reader?] should cultivate a sensitivity both to the uniqueness, importance, and complexity of individual detail . . . and to the network of relations—what he called "constellations"—in which they are embedded, drawn together, and made meaningful.*
>
> —Dick Hebdige

1.

This is a book about resistance in the texts in English of a select group of primarily literary writers, most of whom are from the formerly colonized countries of the Caribbean, Africa, and South Asia and who now reside (or have resided for extended periods of time) in the West—mainly Britain, but also, in a couple of instances, the United States. Specifically, the writers whose work this book focuses on are C. L. R. James, Salman Rushdie, Ama Ata Aidoo, Michelle Cliff, and Hanif Kureishi. At first glance what might seem a curious mix of writers will not, I hope, seem so curious after all once I have had the opportunity, in this introduction and the chapters that follow, to demonstrate significant affinities in their positions on, and enactments of, resistance, which accrue, in large part, from their locations within the metropole.[1] Residence within the metropole, I argue, is an enabling condition of the work of all five writers. Or to out it another way, the metropole is the site that has generated not only Western/metropolitan knowledge(s) about the colonial others, but also, as I seek to demonstrate, the negotiations—including overt contestations and acute modifications—of these knowledge(s) by these writers from the so-called colonial peripheries, now situated "at the heart," as Antoinette Burton entitles her work alluding to a somewhat different context "of [a former colonial] empire." Adapting from Douglas Haynes and Gyan Prakash's definition of resistance in their introduction to *Contesting Power,* by resistance I mean, in particular forms of "cultural [/textual] practice[s] by subordinate [i.e., formerly colonized] groups [and individuals] . . . contest[ing]" the hegemony of their former colonizers, Britain, in particular, and Europe, in general.

The primary audience of these writers' texts is also situated in the West. But insofar as these texts also circulate in the (formerly) colonized spaces of the Caribbean, Africa, and the Indian subcontinent, these texts address audiences in these spaces as well.

Through close readings of selected texts by these writers, this book seeks to enlarge or, at the very least, to modify our understanding of the subject by studying diverse strategies and enactments of resistance as products of specific historical circumstances and rhetorical exigencies and their modes of recuperation. This entails attending as well to those mediations—of ideology, theoretical and other assumptions, desires and expectations—that underpin not only the ways in which these writers conceptualize resistance but also how we as readers recuperate their conceptualizations.

The inquiry from which this book has emerged itself arose from a sense of unease and dissatisfaction based on two interrelated sources: 1) the recognition of a lack of fit sometimes between what a writer conceived of as his or her counterhegemonic practice and a recuperation of this practice via a notion of resistance at odds with this conception such that it was rendered invalid or, most often, ignored altogether; 2) the recognition, more generally, that most discussions of resistance, in the process, perhaps, of authorizing their views, were framed to exclude (even disparage) those conceptions of resistance that ran counter to their own emphases. For example, in courses in which I have taught C. L. R. James's *The Black Jacobins* (1938) and *Beyond a Boundary* (1963), his deep embeddedness in and invocations of his "Englishness" have proven to be deterrents to any serious consideration of his revolutionary credentials as a theorist of (and actor, in some cases, in) decolonization movements in the Caribbean and Africa. In a related fashion, a focus on the predominantly positive reception accorded to Rushdie's work in the West has led readers like Timothy Brennan, for instance, to construe Rushdie as a "cosmopolitan" writer whose literary "eye" is fixed on, and a product of his (i.e., Rushdie's) efforts to satisfy the expectations of the metropolitan publishing industry and readership. While proffering a valuable framework within which the "methods of cosmopolitanism" can be evaluated, such a perspective nonetheless gives short shrift to Rushdie's own claims about his subversive positioning, which, while not at odds with his "cosmopolitanism," cannot simply be reduced to it either.

Deploying a different point of departure, I am not so much interested in contesting such views as I am in providing a different angle of vision through which resistance in the texts of writers like James and Rushdie can also be conceptualized. Once we have acknowledged and engaged with the impact of Western epistemologies, including their "technolog[ies] of cultural meaning" (which is how Burton aptly defines representations [8]), is it possible, I ask, to discern counterhegemonic knowledge(s) and procedures at work in these texts? My analysis argues that it is not only possible but also necessary, certainly if we are to give credence to these writers' own assessments about their practice.

There is a third, arguably just as important impetus for this study. At the time it was conceived, the more substantive and valuable treatments of resistance, revolving around questions of definition and interpretation, and theorizing on the basis of specific historically situated examples, tended to appear in the human and social sciences—anthropology, history, politics.[2] (In section 2 of this introduction, I will have occasion to examine, in some detail, select examples of this work.) Given that "Third World" literatures were being (and still are) in the process of being incorporated within English department curricula in metropolitan universities, and given, moreover, that resistance, as I will demonstrate shortly, was (and still is) a prominent rubric under which materials from the "Third World" were (and are) being studied, there was for those of us who teach these literatures a pedagogical need for bringing into our analyses of these literatures the insights generated by the work on resistance in the human and social sciences. My book attempts to address this need.

But to further clarify the parameters of this study, why it was undertaken and what it seeks to accomplish, a more extended account of its initial impetus is necessary.

2.

When I was hired to teach "anglophone literatures of the 'Third World'" at Oberlin College, I outlined (in the syllabus for what has come to function as a "core" course in these literatures) what I then perceived as key issues or questions pertaining to the difficulties in defining what was then a quite "new" field of knowledge.[3] Included in the outline were questions about the usefulness of categories like the "Third World," of concepts like "postcoloniality," "marginality," and so forth. Since the course was heavily weighted toward a study of novels, many of them in the "realist" vein, I raised questions pertaining to the ideology of this genre. And I asked whether or not these texts, emerging from and/or about the formerly colonized anglophone "Third World," could be usefully studied as enactments of textual resistance as I have defined it above.

Not surprisingly, resistance was the preferred topic for our discussions, found capacious enough not only to include the means by which all the works we were studying could be discussed, but also to subsume the other questions I had raised within its folds. Nor was this an idiosyncratic choice on the part of my students. For if we look at the by now fairly extensive scholarship, in a variety of disciplines, on what are defined as "Third World" (or "postcolonial") subjects, resistance—as subject matter, strategy, even a way of being—figures prominently in this scholarship.[4]

What I did find somewhat disconcerting at first, however, was the sharpness, clarity, and ease with which my students felt they could distinguish between writers, works, and cultural practices they found resistant and those they did not. Moreover, the sharpness and clarity of their distinctions, with few exceptions, depended on or was underpinned by a stark dichotomy that pitted (a predatory, dominant) "West" against (a subordinate, victimized) non-Western "Third World." Any writer, work, or cultural practice that did not aggressively confront and reject the "West" or anything "Western" was declared a witting or unwitting accomplice of the "West"—deemed enthralled by and thus collusive with its value systems. By definition, then, such a writer, work, or cultural practice could not be characterized as resistant.[5] Concomitantly, those writers, works, or cultural practices that declared, mostly in a vocabulary of cultural nationalism, their unremitting opposition to everything Western were, again by definition, found resistant. As my students elaborated upon their judgments, it was hard not to be struck by the irony of these judgments being made in a privileged corner of the West they excoriated. It was also hard not to be struck by the fact that *all* the works we were reading were written in English, by writers from the middle and upper classes, whose incorporation, no matter how selectively achieved, into Western ways of thinking and being could not be so easily gainsaid.

The desire for purity, for a resistant subject and subject matter somehow uncontaminated and therefore uncompromised by (or despite) its encounter with the West, which fueled my students' assessments of what counts as resistance, is not restricted to my students alone. It haunts a fairly substantial and intellectually respectable strand in analyses of resistance in a number of essays and books on the "Third World." In these essays and books, for example, the dichotomy on which my students' judgments were based is also in evidence, albeit in a more attenuated form, because it is framed by more learned and theoretically informed scholarship and sometimes couched within appropriate disclaimers. To mention but two recent examples of such work, both from literary-cultural studies, this dichotomy underpins as insightful and provocative a study as Tim Brennan's on "Third World" "cosmopolitan" writers, as, too, it underpins Benita Parry's impressive and rigorous defense of anti-colonial nationalisms.[6]

Investigating the conjuncture between negative assessments of all nationalisms in the West, particularly by prominent Western scholars, and the positive reception accorded, again in the West, to "cosmopolitans and celebrities" like Salman Rushdie from the "Third World," who also exemplify "a dismissive and parodic attitude towards the project of national culture," Brennan sharply distinguishes "cosmopolitanism" from "what Barbara Harlow . . . calls 'resistance literature,' the literature of independence movements" ("Cosmopolitans and Celebrities" 7, 4). In "cosmopolitanism," says Brennan, an "artistic embrace of the West" is accompanied by a "rejection of the caustic mentality of resistance" ("Cosmopolitans and Celebrities" 4).[7] He restates this distinction between a Western-oriented (and nurtured) "cosmopolitanism" and authentic forms of "Third World" resistance when, defining the former as "involv[ing] not so much an elite at home, as it does spokespersons for a kind of perennial immigration, valorized by a rhetoric of wandering," Brennan asserts that "cosmopolitanism," by definition almost, "violates an important Third World mode," which, in his critical exegesis on "cosmopolitans and celebrities," he links integrally with a commitment to "Third World" national(ist) agendas ("Cosmopolitans and Celebrities" 2, 10). Brennan's strong critique of "cosmopolitanism," underwritten by the binarism the West versus the rest, is valuable precisely because it implicitly urges us to interrogate critically those discussions of "migrancy" and diasporic position(ings) in postcolonial and cultural studies that see resistance often only as a function of migrant or diasporic locations within the West. Additionally, though my study does not view resistance as a function only (or even primarily) of migrant or diasporic locations, insofar as it does focus on the resistances we can discern in the texts of migrant writers in the metropole, Brennan's critique offers an important counterpoint to a study like mine. Juxtaposed with each other, they can help define the (inevitable) limits of each other.

Like Brennan, Parry, too, finds in anti-colonial ("Third World") nationalisms and nativist positions and formulations important sites for resistance. Somewhat differently from Brennan, however, the objects of her critique are those theories of "colonial discourse" (like Gayatri Chakravorty Spivak's and Homi Bhabha's, for instance) that, drawing upon and exemplifying deconstructive and poststructuralist methods, blur or render ambivalent the boundaries between domination and resistance, colonizer and colonized. In so doing, they displace, even obscure what Parry, quoting from Frantz Fanon, characterizes as the "'murderous and decisive struggle between two protagonists" ("Problems" 43).[8] Emphasizing conflict and antagonism, with the colonizer and colonized functioning as "hostile interlocutors" ("Signs" 12) and implacable opponents, Parry locates resistance in "another knowledge and alternative traditions" ("other" and "alternative," that is, to the colonizer's), which are the possessions of the "[colonized] native as historical subject and combatant" ("Problems" 34).[9]

Of course, there is also an extensive body of work on resistances enacted by the oppressed, the marginalized, and the disenfranchised in the context of anti-colonial and other struggles against domination, which refuses a singular or clear-cut model on which resistance is (or can be) based and/or evaluated. (This body of work, incidentally, constitutes the privileged mode of analysis in studies of resistance, unlike Brennan's and Parry's, for example, which constitutes a "minority" strand, as it were.) Additionally, in a number of instances, this work, on principled and pragmatic grounds like Brennan's and Parry's, refuses to draw the lines clearly and decisively between domination and resistance, colonizer and colonized. Heterogeneous in terms of its points of departure, theoretical assumptions, and ideological investments and agendas, this work underscores "the difficulty of situating resistance" (Sharpe 147), not the least because the meaning of a resistant act, or practice, or event, is not transparent or simply "given"; instead, as James Scott points out, "it is socially constructed" (205), a product of interpretation in which historical moment and milieu interact with the purposes and interests of those assigning meaning. Eschewing the tendency to focus on confrontation and overt trials of strength as constitutive elements of resistance, this work has expanded the scope of what counts as resistance by taking on board a whole range of humbler and less obvious or visible strategies and processes through which sociopolitical and cultural transformations are accomplished.

Jean Comaroff, for example, in her study of tribal life in South Africa, urges us not "to discount," as we are likely to if we focus only on "revolutionary success," "the vast proportion of social action . . . played out on a humbler scale." For if we do so, she adds, "we also avoid, by teleological reasoning, the real question as to what are the transformative motors of history" (quoted in O'Hanlon, "Recovering the Subject" 215). In a similar vein, Stuart Hall, while noting that the "classic metaphors of transformation are modeled on 'the revolutionary moment,'" points out how the contributions to *Resistance Through Ritual* (a volume of essays on "Youth

Subcultures in Post-war Britain," published by the Birmingham Centre for Contemporary Cultural Studies, of which Hall was a director) "actively distanced [themselves] from the classical metaphors" because these metaphors were inadequate in accounting for those "forms of disaffiliation . . . which were in some sense challenges to and negotiations of the dominant order but which could not be assimilated to the traditional categories of revolutionary class struggle." By proffering instead of the "reform/revolution antinomies" analysis that drew upon the "Gramscian notion of 'repertoires of resistance,' which [the essays] insisted were always conjunturally defined and historically specific," *Resistance Through Ritual* also proffered "an expanded definition of social rupture" through which transformations within a society are produced ("For Allon White" 294).

In another influential model for studying resistance, James Scott's *Domination and the Arts of Resistance: Hidden Transcripts* argues that there is a range in the strategies and political discourse subordinate groups deploy in their interactions with those who dominate and have power over them. Among these, he scrutinizes the modalities represented, on the one hand, by the "public transcript," which describes the "public performance of the subordinate [that, "with rare . . . exceptions,"] will out of prudence, fear, and the desire to curry favor, be shaped to appeal to the expectations of the powerful," and is therefore "accommodationist" in tone (2, 3), and, on the other, the "hidden transcript," which "characterize[s] discourse that takes place off-stage, beyond the direct observation by powerholders" (3). He also scrutinizes a range of practices located between these two. Through the "hidden transcript" a "sharply dissonant culture is possible" (15), but typically the "hidden transcript" "consists of off-stage speeches, gestures, and practices that confirm, contradict, or inflect what appears in the public transcript" (3). Noting that the "hidden transcript . . . is specific to a given social site and . . . elaborated among a restricted 'public,'" Scott also expands the definition of resistance(s) it enables to include a "whole range of practices . . . such as poaching, pilfering, clandestine tax evasion, and intentionally shabby work for the landlord" by peasants (14). These practices, easy to disregard in (or not even visible to) analyses that privilege open confrontation or direct challenges to domination, are here brought within the purview of what might also constitute resistance via attention to local, small-scale, and humbler actions through which subordinate groups manage to "imply that they are grudging conscripts to the performances" elicited by the "public transcript" (15).[10]

Like Scott, for whom the "frontier between the public and hidden transcripts is a zone of constant struggle between dominant and subordinate—not a solid wall" (14), the cultural-materialist analysis of theorists like Alan Sinfield and Jonathan Dollimore draws upon the insights of Raymond

Williams to argue "the co-occurrence of subordinate, alternative, oppositional cultural forms alongside the dominant, in varying relations of incorporation, negotiation, and resistance" (Sinfield 9). Thus, both Sinfield and Dollimore urge assiduous attention to those moments of "conflict and contradiction that the social order inevitably produces within itself, even as it attempts to sustain itself," because it is in these moments that we can discern the "inability or refusal" on the part of the subordinate "to identify . . . with [the interests of] the dominant, and hence where dissidence may arise" (Sinfield 41–42). For Dollimore (whose *Sexual Dissidence* examines a range of resistant, or rather, dissident practices from attempts by the subordinate at inclusion in the dominant to deviant identifications, such as inversions or perversions, by the subordinate with the dominant's cherished values, beliefs, and self-descriptions), because "contradictions render social processes the site for contest, struggle, and change," making contradictions visible "contributes to transgressive or dissident knowledge" (88). This work in cultural materialism, incidentally, has also added to the vocabulary through which a range of resistance practices can be specified—dissident/dissidence, subvert/subversion, transgress/transgression—and whose meanings extend from disagreement and dissent to the more potent overthrow or ruin or destruction of something established, like a government or institution or belief.

Resistances enacted by women have provided especially fertile ground for theorizing "unlikely forms of resistance"—resistances that are "subversions," as Lila Abu-Lughod puts it, "rather than large-scale collective insurrections, small or local resistances not tied to the overthrow of systems or even ideologies of emancipation (41). In an account that marries, as it were, the insights of feminist theor(ies) with those of Scott's and others on the more pervasive, "everyday" forms of resistance, Abu-Lughod's "The Romance of Resistance" looks at "the changing situation of women in a Bedouin community in Eygpt's western desert." Employing Foucault's formulations about power and resistance as her starting point, Abu-Lughod's interest in using "resistance as a *diagnostic* of power" leads her to what "the rich and sometimes contradictory details of resistance" by these women tell us about "the workings of social power (42). The Bedouin women she interviewed and writes about in this essay, she says, both collude with and resist their confinement within a gender ideology "enforced by the elder men of their community" by "enact[ing] all sorts of minor defiances" that are yet practiced within the parameters of what is defined as their space in a "system [underwritten by patriarchal ideologies] of sexual segregation"; but they make this space an enabling one, "fiercely protect[ing] the inviolability of [this] separate sphere" because this is the arena within which their "defiances can take place" (43).[11]

A majority of the accounts that look for and discern resistances in unlikely places focus as well on the complexities and ambiguities in motivations and processes leading up to the performances of resistance. Additionally, arguing, as does Sherry Ortner, for example, that "individual acts of resistance, as well as large-scale resistance movements, are often themselves conflicted, internally contradictory, and effectively ambivalent in large part because of these internal complexities" (287), scholars contributing to our understanding of these forms of resistance argue that the forms of domination that provoke these individual and collective acts of resistance are also themselves marked by contradictions and ambiguities and are therefore not exempt from internal instabilities and incoherence. Included among these accounts of contradictions and ambiguities within resistance and domination are analyses that address what Simon Gikandi, in a somewhat different context, characterizes as "the mutual imbrication and contamination" of the dominant and the subordinate, colonizer and colonized (xviii), which in turn renders each "impure," or, as Stuart Hall puts it, "inextricably mixed" and "hybrid" ("For Allon White" 292). Thus, we are reminded again and again that the categories through which we set out to examine resistance (and domination), should be flexible, "multifarious and nuanced" (Rosalind O'Hanlon, "Recovering the Subject" 281).[12]

The studies of resistance that addresses the "mutual imbrication" of the discourses of the dominant and the subordinate are able to better (and more persuasively) account for the forms of resistance we can discern in literary texts, as I point out in my opening remarks, that are most often taught in courses on the "Third World" in metropolitan universities. Primarily in English, these texts about the "Third World" also constitute one of the predominant ways in which the "Third World" is made available and consumed in the West. I should also add that insofar as such understandings of resistance speak more accurately to my own experience and cultural and intellectual formation, I often tend in my own courses to choose several literary texts that foreground mixtures and impurities in practices and/or strategies of resistance. (Examples of these literary texts include: Merle Hodge's *Crick-Crack Monkey,* Tsi Tsi Dangarembga's *Nervous Conditions,* Bessie Head's *A Question of Power,* Earl Lovelace's *The Wine of Astonishment,* Tayib Salih's *Season of Migration to the North* (originally published in Arabic but now easily available in a fine English translation) and Salman Rushdie's *Midnight's Children, Shame,* and *The Moor's Last Sigh*. Incidentally, a number of these texts address the "mixed blessing" that colonial education turns out to be.) For as a recipient of what is often thought of as a classic instance of (neo-)colonial education (I was educated in a convent run by Irish nuns, with a college degree in English literature that focused exclusively on canonical English texts), I have found that my cultural and intellectual formation is similar, if not identical, to that of writers whose texts I teach.[13]

However, not all writers from the "Third World," despite their entanglements—which have to do as much with their education as with their location—within the discourses and values of the West, position themselves in identical ways. Some, like Ama Ata Aidoo (and Michelle Cliff to a somewhat different degree), are invested in a more starkly oppositional stance and rhetoric through which they distinguish (and urge their readers to distinguish) clearly and decisively between the subjects who resist from the objects of their resistance. Their positions are important to map (as I do in my courses as well as in this study) for at least three interrelated reasons: (1) Insofar as these are positions that make sense to, and possess a "reality" for, some "thoughtful" minds, they constitute "phenomena to be accounted for,"[14] not something to be ignored or consigned to irrelevance. (2) In a study that assumes that the mediations of ideology (and the ways in which these mediations affect one's desires and expectations) all constitute phenomena through which particular perceptions of resistance can be accounted for, surely, starkly oppositional positions on resistance, no less than those that foreground conflict, contradiction, and ambiguity in the enactments of resistance, provide important "material" resources. (3) The *difference*—in the strategies they emphasize, in their affective investments, even in the vocabulary through which they express their claims and judgments—between starkly oppositional ways of thinking about resistant practices and positions and conceptions of resistance that stress contradictions and complexities can help us see the limitations of each and thus provide us with a way for thinking critically about each. It is in my desire to honor, but also to reflect critically on, these divergent—opposing, some would argue—positions on and ways of accounting for resistance that I locate one of the major purposes of this study.

An important disclaimer about the *difference* I am invoking above is in order here. Critics have often claimed that no modes of resistance, whether they acknowledge it or not, are completely free of their implication in the domination they resist. Thus, Richard Terdiman, in his study of the "theory and practice of symbolic resistance in nineteenth-century France," contends: "Counter-discourses are always inter-locked with the domination they contest"; "counter-discourses," he adds, "are never sovereign. And they are least so when sovereignity is the claim they most seek" (11). Similarly, Sarat Maharaj, in remarks I will have occasion to quote several times in the course of this book, notes that although "the drive behind the oppositional mode seems to be the desire to find some stance *outside* [the dominant]," this stance "is not so much found ready-made 'out there' as it has to be painfully constructed from *within* . . . out of the elements of the very system of [domination] it seeks to go beyond" (15; emphasis added). Something comparable can be asserted on behalf of those accounts that locate "true" resistance in uncontaminated and implacable opposition to the dominant. As I demonstrate in the second section of this study, the use of inversion, for example, which is a favored strategy in the arsenal of this form of resistance, would not be possible without the terms and evaluations embodied in and by the dominant, which the inversion then attempts to devalue (and revalue) through a process of transvaluation. But the *effect* of these accounts, I would argue, is different from those that overtly acknowledge the "mutual imbrication" of the resistant with the dominant modes and practices. This *effect*—and my argument in this regard applies to all forms and accounts of resistance and their effects—is produced not only by the way in which a given form of resistance is enacted and described but also by the fact that this enactment and description are themselves products of specific circumstances—historical, political, cultural, and social—and assumptions regarding the resisting subject's "subjectivity, identity, and agency" (Parry, "Resistance Theory" 172), and finally the ways in which the actors and transmitters of this resistance position themselves with respect to a given form of resistance. Taking these into account entails attending to the differences they embody and in turn produce.

In this study, I demonstrate the differences and overlaps between the strategies and practices of resistance I have outlined above by dividing my argument into sections—three altogether—and then juxtaposing these sections with each other so that they interact with and throw into relief the alternative, or even "opposing" conception of resistance that I formulate in each section. The first two sections, in turn, contain two chapters each, with each devoted to the work of a single writer—C. L. R. James and Salman Rushdie in the first, and Ama Ata Aidoo and Michelle Cliff in the second; the third section contains only one chapter, again devoted to a single writer, Hanif Kureishi.

While the writers and the texts I have chosen for close analysis are by no means the only ones through whom an analysis like mine on resistance can be undertaken, they have served as my primary examples for some time now in the courses I teach and the research projects I've pursued. Thus, extensive familiarity with these writers and their texts, aligned to what I view as their ability to speak cogently and well to the positions I was exploring in this study, made them the most useful choices for my purposes. Additionally, occupying varying geographical locations and somewhat different historical moments, and exemplifying a range of political sympathies, they provide me as well with appropriate grounds for the kind of comparative and contextualized analysis I wished to carry out in this study.

With reference, for example, to range in terms of geographical locations inhabited prior to, and/or in conjunction with, these five writers' residence in the metropole, consider that James and Cliff come from the Caribbean (Trinidad and Jamaica respectively), Aidoo from Ghana, Rushdie from the Indian subcontinent, and Kureishi from south London. Most of them travel back and forth between their places of "origin" and of residence. Although all but one of the five writers are alive and writing today (James died in 1989), there's some range as well in the historical moments they were born into, with James coming first (1901), then Aidoo (1942), Cliff (1946), Rushdie (1947), and Kureishi (1954). Similarly, there is some range in the historical moments of the texts I examine closely, with James's *The Black Jacobins* (1938) and *Beyond a Boundary* (1963) on one end and Kureishi's *The Buddha of Suburbia* (1990) and *The Black Album* (1995) on the other framing, as it were, a selection that extends from Aidoo's *No Sweetness Here* (1970) and *Our Sister Killjoy* (1977) to Rushdie's *Midnight's Children* (1981) and *Shame* (1983) to Cliff's *Abeng* (1984) and *No Telephone to Heaven* (1988). With regard, then, to the constituting and constitutive forces of the colonial encounter between Britain and its (ex-)colonial possessions, which is one of the major preoccupations of all five writers (and the texts) I examine, despite affinities that accrue from their class affiliation as members of the educated middle to upper middle class and from an English (colonial) education, each writer is nonetheless also situated differently in terms of her or his relationship to colonialism. In this respect, it's also important to note that whereas James, Cliff, and Aidoo lived or have lived through, even if minimally, times when colonial rulers were still in power in the countries in which they were born (Trinidad became independent on August 31, 1962; Jamaica on August 6, 1962; and Ghana on March 6, 1957), Rushdie, born in June 1947, a mere two months before the British formally departed from the Indian subcontinent and Kureishi born in 1954 were born in allegedly "post" colonial nations.

3.

What ties these writers together (and influences significantly the work by them that I have selected for close analysis) is their location—their residence in the (imperial) metropole (primarily Britain, but also the United States in some cases) as members of diasporas from former British colonies in the Caribbean, Africa, and the Indian subcontinent. I use the term "diaspora" with some trepidation, not the least because it has acquired a sort of "modish sheen" (Said's phrase), a privileged place in discussions of "Third World" subjects in metropolitan institutions—a modishness and privileging my study undoubtedly taps into to some extent. However, given that this "term," as Khachig Tölölyan puts it, which "once described Jewish, Greek, and Armenian dispersion, now shares meanings with a larger semantic domain that includes words like immigrant, expatriate, refugee, guest-worker, exile community, overseas community" (4–5), its "loosely coherent" (James Clifford, "Diasporas" 310) applicability to these "*different categories* of people" (Cohen 21; original emphasis) is particularly useful for my purposes.

Among the characteristics associated with a diaspora, those central to my analysis deal primarily with what Clifford identifies as "diasporic forms of longing, memory, and (dis)identification . . . shared by a broad spectrum of minority and migrant populations" within the (imperial) metropole ("Diasporas" 304). Drawing upon William Safran's definition of a diaspora, Clifford's reference to "longing" and "memory" derives from Safran's assertion that "expatriate minority communities" that constitute a diaspora are deeply invested in "a memory, vision, or myth about the original homeland," to which they hope to return (83), although, as Clifford and others qualify, the "teleology of 'return'" is more often than not symbolic—achieved via the imagination or imaginative constructs—rather than literal.[15]

The "(dis)identification" (with the host country) that Clifford alludes to has at least as much to do with a diasporic peoples' belief that "they are not—and perhaps cannot be—fully accepted by their host" (Safran 83) as it has to do with an effort to reclaim "another nation that has been lost, elsewhere in space and time, but [continues to be] powerful as a political formation here and now," especially given these peoples' displacement from a prior "home" and within an inhospitable space marked by "structural prejudice" (Clifford, "Diasporas" 307).

Both Cohen and Clifford point to the diaspora as the site for the "construction of . . . new identities and subjectivities" (Cohen 128) and of "alternate public spheres, forms of community consciousness and solidarity that maintain identifications *outside* the national time/space [of the host] in order to live *inside, with a difference* (Clifford, "Diasporas" 308; emphasis added). This "difference"—a product of diasporic peoples' "(dis)identification" with the host—as well as the "new identities and subjectivities" generated by the diaspora through mixtures and hybridization, as its peoples simultaneously incorporate and resist the hegemonic values of the host, can become important instruments in the political struggle to redefine the dominant—both their relationship to the dominant as well as the culture and power the dominant represents.

Of course, the writers I examine are positioned (and/or position themselves) differentially with respect to the characteristics of a diaspora I have outlined above. Both James and Rushdie, for instance, purposefully invoke their liminal—insider/outsider—location as a form of "(dis)identification" that empowers their critique of hegemonic Western/colonial ideology: "*In* the West, but not [necessarily] *of* it," as Paul Gilroy puts it.[16] (Rushdie extends this positioning to incorporate his relationship with his "homes"—India and Pakistan—as well. *Shame,* for example, is a narrative of nation explicitly constructed from the perspective of an immigrant who has "learned Pakistan in slices" [70], but whose "long geographical perspective" ["Indian Writer in England" 79] is the basis for this alternative, critical account.) For Kureishi, a native-born Briton who only partially belongs to the South Asian diaspora in that his father is an immigrant to Britain from Pakistan but his mother is English, nonetheless deploys his position as a "hybrid" to anchor his argument about "a new way of being British" ("The Rainbow Sign" 37).

Conversely, Aidoo's *Our Sister Killjoy* can be read as a pointed polemic against the traps of migrancy and "self-exile." The lure of "dwelling-in-displacement" (Clifford, "Diasporas" 310) that a diaspora represents is for her an imperialist weapon of division ("An enemy has thrown a huge boulder across our path. We have been scattered. We wander too far. We are in danger of getting completely lost. We must not allow this to happen" [*Our Sister Killjoy* 118]). Moreover, diaspora dwellers represent the most exacerbated forms of deracination from which she seeks to redeem them via her pro-return-to-Africa, anti-migration-to-the-West argument. The African diaspora within the metropole, however, is her crucial referent, the most significant foil to her nativism. What's more, *Our Sister Killjoy* came together for her and was completed during one of her extended stays in the United States.

In a somewhat different vein, but with a similar end in mind, Clare Savage, the autobiographically informed protagonist of Michelle Cliff's *Abeng* and *No Telephone to Heaven,* comes to a critical self-awareness of her political commitments via her extended stay in the metropole. Her return to Jamaica and to the "African part of [herself]" (*The Land of Look Behind* 14) is routed through and embraced during her displacement, first in the United States and then more substantially in Britain. Cliff, of course, lives in the United States and "wonders if [she] will ever return," but "keep[s] track of [her homeland] through advertisements and photographs . . . a group of apple pickers . . . shots of graffitti on the walls of Tivoli Garden" (*The Land of Look Behind* 103).

James's anti-colonial vision was also forged in the metropole, as my close readings of *Beyond a Boundary* and *The Black Jacobins* make clear; similarly, it was the instruments of empire—cricket and the Enlightenment narrative exemplified by the French Revolution—that were appropriated, with a difference, for his efforts on behalf of decolonization. Nostalgia of the kind that infuses Cliff's "memory" of and "longing" for Jamaica, however, have little significance for James, whose *The Black Jacobins* was fueled by the practical intent to "stimulate the coming emancipation of Africa" ("Preface" to the second edition).

Certainly, nostalgia seems to have little or no place in Rushdie's symbolic "returns" to his "homeland(s)"—India and Pakistan—via his (alternate) narratives of nation in *Midnight's Children* and *Shame,* which, as I point out in my chapter on Rushdie, are acerbic critiques of what these nations have become. And yet, when he describes in "The Indian Writer in England" the genesis of *Midnight's Children,* it is precisely "memory" and "longing" fused together in a heady mix that fundamentally inform his desire to reclaim Bombay/India. Revisiting the house in which he was born in Bombay, Rushdie observes: "Bombay is a city built by foreigners upon reclaimed land. I, who had been away so long that I almost qualified for the title of *farangi* [foreigner], was gripped by the conviction that I, too, had a city and a history to reclaim" (76).

Somewhat differently, Pakistan, which is for Kureishi not quite a prior "home," but one that nonetheless "contain[s] fragments of [his] past," "stories to help [him] see [his] place in the world," provides him with an alternate understanding of Britain—of what Britain should be; Pakistan, furthermore, enables him to "form an image of [himself]" that he can "take back . . . to England" to use as a form of "protect[ion]," a resource for strength ("The Rainbow Sign" 33).

One of the means through which the "homes" left behind are reclaimed in a diaspora is via reconstructions of their histories/historical and national identities. Not once mentioned in the theories of diaspora forwarded by Clifford, Cohen, and Safran, these reconstructions by Cliff and Rushdie nonetheless constitute an important way in which the "nation that has been lost, elsewhere in space and time" comes to function as a "powerful political formation [in] the here and now" of the metropole. Alternative histories of their "homelands" undertaken by minority/(formerly) colonized groups and individuals who wish to speak from their perspectives and in their own interests, these reconstructions contest and/or call into question the truth of dominant, colonial histories of their "homelands" (although in Rushdie the history he contests also includes the dominant nationalist one). Barring Kureishi, each one of the writers I examine engages explicitly with the task of historical retrieval and reconstruction. This revisionary task, as I demonstrate, is carried out differently by each of the writers I examine, issuing sometimes as overt contestations of dominant versions, sometimes as excavations of buried or deliberately suppressed histories that call into question dominant accounts. Some of these revisionary versions insist on their truth over that of the dominant version's; some simply range their insights alongside those of the dominant version's, offering what they conceive of as alternative accounts.

4.

Although the elements that characterize a diaspora provide one of the more significant contexts within which I situate the writers and texts I study, there are other contextual elements to which I attend in the course of my analyses. In addition, then, to the (diasporic) location of each writer and the personal and political affiliations that are in great measure a part of this location, I attend to the historical moment and milieu from which each writer and his or her work emerges; the genre to which a given work belongs, even (or especially) when it seeks to rewrite that genre; the audience(s) each writer tries to address through her or his work; and finally the ways in which each writer (and work) positions herself or himself with regard to gender in her or his account(ing)s of resistance, particularly because at least four of the writers I have selected for close study explicitly link the workings of patriarchal ideologies with the dominant and those of their presumed obverse with resistance, so that "woman" and the place "she" occupies is envisaged as the site wherein counterhegemonic values and stances are nurtured and/or located. I find these elements of historical moment and context crucial to any analysis of resistance (including mine) not only because they constitute the mediations affecting the ways in which resistance is conceived of, performed, or described, but also because these mediations affect and therefore enable us to apprehend the possibilities available to, and/or the constraints placed on, a given subject or form of resistance.

It is important at this juncture to clarify that these details of historical moment and context, which I consider so important to our study of resistance, are seldom simply available "out there" to be retrieved and slotted into our analyses. As I have been intermittently suggesting in this introduction, this retrieval (or, more precisely, those elements that we do retrieve) is itself a product of a particular reading and hence of particular agendas and investments, including pre-set categories for evaluating what counts as resistance. I do think, however, that writers *intend* specific effects and seek to forward certain agendas, and though these can be arrived at only through a process of reading and interpretation, careful and assiduous attention to context can (and to my mind does) operate as "'the setting of limits'" and the "'exertion of pressure'"—on our readings—which Gauri Viswanathan suggests is "closely allied with" what Raymond Williams identifies as "the determinative principle in intention" (54). It is with a view to attending to these "limits" and "pressures" that I allot considerable space to what the writers have to say about positions that have a bearing on those exemplified in the texts I look at. While I do not expect my readers to forget that I am the one carrying out the readings that constitute this study, I hope that they will be persuaded, on the basis of the "evidence" I muster with regard to historical context and moment and other contextual elements, that these readings could have been intended (in some of the ways specified above) by the writers themselves.

There's a circular rather than linear, or even developmental, shape to this book, in which I begin and end with somewhat differing conceptions of resistance that proceed from inhabiting the dominant or hegemonic forms of belonging, which in this instance, since we are dealing with the work of immigrants from former colonies to the colonial metropole, translates to the discourses and values primarily of the West. The section that constitutes the middle examines writers and texts that locate themselves squarely on the terrain of anti-colonial (cultural) nationalism, which, in privileging an absolute division between the colonizer and colonized—in terms of their representations, values, and so on—operates in a contradictory or contentious relationship with the sections that precede and follow this one. This arrangement, whereby the starkly oppositional conceptions of resistance are framed by sections that examine resistances which recognize their co-implication with the dominant or hegemonic, might well be seen as an effort to contain, even dilute the force of, forms and enactments of resistance that seem to run counter to those this study seems to privilege by virtue of granting them more space in the analysis. I hope that this is not the case, for through this framing I am not so much recording a preference (although I am doing that too, but minimally) as I am, through structural and other means, recording my recognition that insofar as the conceptions of resistance that are examined in the framing sections speak more closely to my ideological formation, they allow me to indicate how this ideological formation undoubtedly has affected my readings in the middle section, as it has affected those in the sections framing it. Despite this recognition, however, I hope it is also evident that here (that is, in the middle section), perhaps more emphatically than elsewhere in this study, I have asked myself again and again questions pertaining to historical moment and context wherein the view of resistance the texts in this section privilege would be not only called for, but necessary.

Section one, "*In,* But Not *of,* the West: Repetition, Appropriation, and Re-vision's of Dominant (Colonial) Knowledges," examines the work of two writers, C. L. R. James and Salman Rushdie, whose resistances—both in terms of their content and strategies—proceed from within the folds of the dominant or hegemonic, appropriating values, concepts, and representations from its "regimes of significations" (Paul Willemen's phrase), especially as these bear on the task of writing history, or of historical reconstruction, which is the task these writers define as central to their revisionary projects. However, because they are differently situated, geographically, historically, and ideologically, the revisionary tasks these writers undertake, despite significant overlaps, are marked as well with different assumptions and goals in mind. James, for example, appropriates from the dominant or hegemonic forms of representation and values in order to transform and extend their scope; he seeks the inclusion of the excluded, marginalized, or oppressed, by way of this appropriation, into the world of, and those values privileged by, dominant or hegemonic forms. Rushdie, on the other hand, seeks to subvert the signifying power and, through that, the authorizing function of the dominant or hegemonic. To this end, his appropriations, which are most often parodic repetitions of dominant or hegemonic systems and structures of representation, are undertaken to undercut these systems' and structures' claim to transcendent truth or value. Regarding this difference between James and Rushdie, we should bear in mind that quite apart from their differing relationships, as people from the Caribbean and the Indian subcontinent, to the metropole, they come out of quite different historical moments as well, which in turn influences their relationships to the projects of decolonization: James, as I've already noted, was born (and lived through a time) when Trinidad was still a British possession. Born before decolonization occurred, he nonetheless oversaw the processes leading to Trinidad's independence. His *The Black Jacobins,* he tells us, was conceived first as a stimulus for "the coming emancipation of Africa"; and then in the Appendix, "From Toussaint L'Ouverture to Fidel Castro," added to the second edition in 1963, he avers that herein he is "attempt[ing] for the future of West Indies, all of them, what was done for Africa in 1938" ("Preface" vii). Rushdie, on the other hand, writing *after* formal decolonization has occurred, and, moreover, when dominant or hegemonic forms include those put in place or embraced by the neo-colonial, nationalist bourgeoisie, subjects both colonial and nationalist representations of the subcontinent to his critique.

In my analysis of *Shame,* I argue that Rushdie's appropriation of a "female" positionality[17] is an important instrument for his counterhegemonic account of the nation—Pakistan. This figurative operation, I suggest, is underwritten by Rushdie's assumption that the domain of the counterhegemonic—as the obverse of the domain of the hegemonic (that is, the "male" domain)—is that of the marginalized and oppressed (that is, the "female" domain).

What Rushdie's *Shame* claims via a figurative operation is rendered literal in the narratives by Ama Ata Aidoo and Michelle Cliff I examine in section two, "Return(s) to the Native Land: The Politics of Cultural Nationalism." Whereas in *Shame* the narrator's appropriation of a "female" positionality is an abstraction that enables his critique of the nation so that his focus remains the "male" narrative, albeit as an object of critique, Aidoo's and Cliff's central preoccupation is with women protagonists and the material conditions they are called upon to inhabit. Their narratives insist that these women, via the positions they take and the commitments they make, enable sociopolitical and historical transformations that are fundamentally linked to the resistances they enact as women. Additionally, though like Rushdie's (and James's), Aidoo's and Cliff's narratives are fueled by critiques of colonialism and neo-colonialism, because of their near-exclusive focus on the damage colonialism and neo-colonialism have wrought—a damage that can be redressed only by making whole what was split apart, the colonized psyche so powerfully delineated by Frantz Fanon as "black skin, white masks"—Aidoo's and Cliff's emphases are different, as is the solution they urge. The "wholeness" both Aidoo and Cliff advocate as a precondition for a genuinely decolonized, self-determining status of their respective nations is premised, I argue, on a literal "return to [their] native land[s]," which, through the logic of cultural nationalism, is in turn linked with an allegiance to blackness that enables, for Aidoo and Cliff, a form of revolutionary praxis and embodies a revolutionary identity. But whereas Aidoo (and her protagonists) assumes an unmediated continuity between her identity and identification with blackness, Cliff (and her protagonists) as a "Jamaican of mostly white ancestry" (Edmondson, "Race" 181) seeks to acquire a black identity through a process of self-education and enlightenment. This difference in the positioning of these two writers also marks the differences in the strategies they adopt and in the particular inflections within their conceptions of resistance.

Aidoo's and Cliff's embrace of nationalism as a form of resistance not only differentiates them from Rushdie and Kureishi, who link nationalism with the perpetuation of dominant (patriarchal) ideologies, but it also differentiates the position they advocate from those feminist critiques that, in locating women's marginalization from and men's centrality to the discourses and practices that authorize nationalism, find nationalism extremely problematic.

It is also instructive that while an important strand in discussions of women's resistances, represented, for example, in Abu-Lughod's work, demonstrates that women typically proceed through indirection and are for the most part nonconfrontional, Aidoo's narrative is cast in sharply antagonistic terms, its deployment of reverse discourse proceeding through a series of implacably oppositional debunkings of metropolitan values and beliefs. Cliff's autobiographically informed protagonist Clare, even as she recognizes the range of positions that fall between those aligned exclusively with white or black identifications as, in fact, positions that describe her own location, nonetheless casts her lot with blackness, which stands for true resistance in her estimation.

The second section, then, not only elaborates a conception of resistance that is substantively different in its emphases from those I analyze in the sections that precede and follow it, but it also asks us to construe women's resistances along somewhat different lines than it has been typically construed.

In the concluding section of my study, "In *and* of the (Imperial) Metropole," I return to the subject of belonging to, and habitation within, the dominant or hegemonic, this time through the example of a writer, Hanif Kureishi, who somewhat differently from either James or Rushdie in the opening section wishes to claim full belonging and the status of an insider ("I am a Britisher," he insists), even as he challenges and critiques dominant or hegemonic conceptions of "Britishness." His claim to full belonging and insider status, however, requires a necessary reform(ulation) of Britishness, whereby "black" and "British" become mutually informing, not mutually excluding identities. Like James (and this is how my argument comes full circle, returning to its point of departure), Kureishi seeks inclusion of the subordinate—the excluded, the marginalized, the oppressed—into the world of, and values privileged by, the dominant.

5.

Wondering about the disquiet among the Tories during the Margaret Thatcher years in Britain, over critiques by liberal and progressive writers of Britain's rundown condition, its aggressive nationalism, and rampant racism, Kureishi is first puzzled and then concludes that these critiques must have hit a nerve; hence the disquiet. "Writing," he declares, "mattered" (*London Kills Me* x). Both Rushdie and Cliff are also convinced of the counterhegemonic function of their (re-)writings of dominant history. Aidoo conceives of her writing as an attempt to address and through that clarify the nature of the significant problems confronting her country and continent. James viewed *The Black Jacobins* as part of an effort to "stimulate the coming emancipation of Africa" and later that of West Indies. Whether the writer links her or his writing directly with a particular struggle, or indirectly by conceiving of it as a way of "enlarging and clarifying human consciousness" that in turn enables its readers to "'see through' and to challenge the conditions that divert living subjects from a real understanding of their interests and their conditions of existence" (D. Miller et al. 13), each writer I discuss in this study sees herself or himself as actively involved in and therefore as producing dissident or counterhegemonic work.

In general, such a claim by writers on behalf of their writing is recognized as valid by others as well. Indeed, Chris Searle asserts a fundamental continuity between writing and different kinds and levels of revolutionary struggle when he says: "For as a people seize a gun and a plough and the other means of production to liberate themselves, they must also seize the *word:* spoken, written, printed, chanted, sung" (*Words Unchained* xxii). Nor is Searle an exception, for as James Scott asserts, "the practices of resistance and discourses of resistance are mutually sustaining" (159). However, although Edward Said, for example, endorses "resistance in the form of critical activity" ("In the Shadow of the West" 6), he nonetheless seeks to discriminate between what he views as different forms, and, for that matter, degrees of oppositional writing. Thus, while discussing Frantz Fanon's *The Wretched of the Earth* and Michel Foucault's *Madness and Civilization* (both published in 1961), he concedes that although both are "oppositional books,"

> It is important that Fanon's book was the result of a *collective* struggle as opposed to Foucault's. . . . Fanon's book is the more powerful because it is rooted in, you might say, the dialectics of struggle. What is present in Fanon's work and absent in the early Foucault is the sense of active commitment. ("In the Shadow of the West" 4)

The distinction articulated by Said—with regard to writing that grows out of direct involvement in a struggle and writing that does not and so, by implication, has a highly mediated or indirect relationship with a given struggle—has been considerably sharpened recently. For example, accounts of textual and/or discursive resistance(s) that draw upon deconstructive and poststructuralist methods emphasizing the fact of mediation are regarded with extreme suspicion as instances of an "exorbitation of discourse" (Parry, "Problems" 43) or as instances, as Ania Loomba quoting from Ann Rosalind Jones puts it, in which "'revolution'" is "reduce[d] . . . to a linguistic and solitary act'" ("Overworlding" 170). Similarly, distinguishing an "activist" from a "textual" culture, Aijaz Ahmad finds that "the increasing dominance of the poststructuralist position has had the effect, in more recent years, of greatly extending the centrality of *reading* as the appropriate form of politics" (3). Further along in his remarks, he derides "very affluent people [who] may have come to believe that they have broken free of imperialism through acts of reading, writing, lecturing, and so forth" (11).

Certainly it is important to distinguish between the political (and ethical) efficacy of what are construed as different forms of resistance/resistant activity. As, too, following Said we can and should distinguish between the political (and ethical) efficacy of different kinds of oppositional writing. But though Ahmad's skepticism with regard to the inflationary claims made on behalf of textual or discursive modes of resistance alerts us to an important limitation in contemporary theory's fascination with textuality, the sharp distinction he sets up between an "activist" versus a "textual" culture is misleading and not very helpful. C. L. R. James, as I've mentioned before, saw his *The Black Jacobins* as providing the stimulus for the projects of decolonization being undertaken in Africa and then the Caribbean. He saw his text, in other words, as leading to—and provoking and nurturing—revolutionary activity. In this view, *The Black Jacobins* is perceived as continuous with, indeed is itself a form of activism. Aidoo, on the other hand, distinguishes on the basis of what each form of activism allows her to achieve, between the kind of activism involved in her writing and that which was involved, say, in her role as education minister for a year in President Jerry Rawlings's government in Ghana in the early 1980s. She accepted the position of education minister, she says in her interview with Adeola James, because "I believe that education is the key to everything. Whereas I do not discount my work as a writer or the possibility of doing things with my writing, I thought that out there as a minister, or whatever, you have direct access to state power, to affect things and to direct them immediately"(11). Significantly, Aidoo wished to use this "direct access to state power" to "shak[e] up, revolutioniz[e]" the "educational structures" (Modebe 40). It's worth noting, however, that though she distinguishes between the kinds of transformations she believes are possible in her work as minister and in her work as a writer, she does not diminish or dismiss the efficacy of the latter whose effects quite simply are differently, more indirectly achieved.

Whether or not we unambiguously accede to claims writers, including the ones I examine in this study, make with respect to the transformations their writing has wrought or can bring about, we would do well to recognize at the very least that these writers claim what they claim in order to write what they do; it is as much a way of specifying a purpose as anything else. Thus their claims ought to be situated pragmatically as much as ethically or politically. In a somewhat similar vein, by offering readings of textual resistance, I am in part, at least, situating myself fairly precisely in terms of my profession and institutional location: Reading and, it is to be hoped, teaching and writing about how to read certain kinds of "Third World" literatures is in large part what I do. Furthermore, my readings in this book are ironically also meant to counter precisely the position Ahmad ascribes to acts of reading undertaken under the aegis of what he characterizes as the limits of poststructuralist theory. Through them I seek to emphasize the mediations that affect not only the enactments of resistance delineated by the writers I examine but also our accounts of them. By so doing, I hope to take some of the moralism and self-righteousness out of our account(ing)s of resistance; for keeping in mind the mediations through which we come to, or that impinge on, our understanding of resistance should prevent us from claiming for ourselves more than we tend to when we treat readings of resistance as if they were a form of resistance itself.

Part I

In, But Not *of*, the West: Repetition, Appropriation, and Re-visions of Dominant (Colonial) Knowledge(s)

In postcoloniality, every metropolitan definition is dislodged. The general mode for the post-colonial is citation, reinscription, rerouting the historical.

—Gayatri Chakravorty Spivak

The anthropologist, [Asmarom] Legesse, has pointed to the extent to which we are trapped in the ordering "categories and prescriptions" of our epistemic orders. He notes, however, that liminal groups of any order are most able to "free us" from these prescriptions, since it is they who existentially experience the "injustice inherent in structure."

—Sylvia Wynter

Chapter 1

Inhabiting the Metropole: C. L. R. James's "Grand Narratives" of Resistance and Emancipation

If the ideas originated in the West Indies it was only in England and in English life and history that I was able to track them down and test them. To establish his own identity, Caliban, after three centuries, must himself pioneer into regions Caesar never knew.

—C. L. R. James

[The excluded, or the subordinate, or the minor] seeks to partake of the dominant term rather than the inferior one. . . . This involves a struggle for inclusion within the very concepts which exclude the subordinate, a struggle which is simultaneously an appropriation and transformation of those concepts.

—Jonathan Dollimore

1.

Despite the call for a "proliferation of historically nuanced theories and strategies"; despite talk about "traditions" as "invented" entities that entail a "selective remembering and forgetting of the past that undergirds group identity"; despite the recognition that histories and identities are constructed; despite the call not to "give in to the rigidity and interdictions of those self-imposed limitations that come with race, moment or milieu; despite (or, perhaps, because of) all these, a significant strand in discussions of anti-colonial resistance, whether in homeland or diaspora, often pits singular notions of dispersed and colonized (African, Indian, or

Caribbean) peoples' national identity and cultural authenticity against the discourses and values that authorize the power of the West.[1] This strand reflects an understandable but also exclusionary desire for resistance to occupy an absolutely autonomous, uncontaminated space from which to launch the "truth" of its "pure" opposition to the West.[2] Drawing upon the power of a stringently oppositional rhetoric, it implicitly or explicitly posits "true" resistance as proceeding from (and only from) identities securely and organically connected to a putative traditional natal culture.

That this view of resistance represents a well-founded reaction to the (continuing) power of Euro-America is obvious. That it requires (indeed demands) "a particularly narrow and specific form [of] dissent" if dissent is to be "audible or politically non-cooptable" (Nandy, "Shamans" 263) should perhaps also be obvious. As such, however, it is inadequate to the task of accounting for those articulations of resistance that emphasize the multilayered, mixed, or hybrid cultural and historical formations such as those produced by and in the face of the massive European colonization of the "non"-West. Instead, then, of writing off as incomplete, inappropriate, or co-opted those modes of resistance that do not adhere to a strictly oppositional stance, these articulations proceed from the recognition that, as Abiola Irele puts it,

> the colonial experience was not an interlude in our [the ex-colonized peoples'] histor[ies], a storm that broke upon us causing damage here and there but leaving us the possibility, after its passing, to pick up the pieces. It marked a sea-change in the historical processes in [the colonized world]; it effected a qualitative reordering of life. (207)

Thus, these articulations suggest, we have much to learn from those resistant visions and voices forged in the crucible of the colonial encounter itself and out of what Cedric Robinson has called "the contradictions of domination, the dialectic of imperialism" ("Amilcar Cabral" 40). C. L. R. James and his work represents one exemplary instance of this take on resistance.

During an interview he gave in the spring of 1974 at Binghamton, New York, C. L. R. James recalls the following anecdote:

> [Aimé] Césaire and I were talking one day, and I asked him: "Where do you come from?" He said, "Well, I grew up in Martinique [and went to] the Victor Schoelcher school" . . . So I said: "What did you do there?" He told me, "Latin and Greek and French literature." And then I said: "What next?" He said, "I went to France, and I went to the Ecole Normale Superieure." . . . And I said: "What did you do there?" And he said: "Latin and Greek and French literature." And then I said: "Where did you go from

> there?" And he said: "I went to the Sorbonne." And I said, "I suppose you did there Latin and Greek and French literature?" And he said: "Exactly. . . . But there is one thing more. . . . I went back to the Victor Schoelcher school, and there I taught Latin and Greek and French literature." So when Césaire wrote his tremendous attack upon Western civilization in *Return to My Native Land* . . . he was able to make this ferocious attack upon Western civilization because he knew it inside out. . . . He had spent some twenty years studying it. (Qtd. in Robinson, *Black Marxism* 259)

What this anecdote captures so well through its sheer repetitiveness, incantatory density, and circularity is the powerful hold of "Western civilization" on young Césaire's mind; however, the circularity is not vicious or self-consuming because, as James tells it, Césaire broke out of its hold when he launched his equally powerful attack on precisely what had held him in its thrall. One could easily argue, as indeed I will argue in this chapter, that in this anecdote James is in fact offering a gloss on his own cultural formation and his method for discerning resistances undertaken by colonized and enslaved populations in the Caribbean; similarly, this anecdote captures the strategies of resistance he himself adopts and/or privileges in his own work.

Like Césaire, James's deep investment in and inside knowledge of the colonizer's culture and world view—English in the case of James—helped him produce his profound critique of English colonialism. With respect to James's social and cultural formation, for example, John Bracey remarks on "a discussion where several comrades and I were railing against Europe and its evils. C. L. R. intervened with 'But my dear Bracey, I am a black European, that is my training and my outlook.' C. L. R. said this without apology, and without seeking our acceptance" (125). Furthermore, as with Césaire, James's investment in the colonizer's culture is as much a matter of his family history and class affiliation as it is of the colonial education that helped him develop his breadth of knowledge about "Western civilization" and that trained him as a "British intellectual" (*Beyond a Boundary* 28, 41, 71, 114). Born in Tunapuna, eight miles from Port of Spain, Trinidad, the son of a father who was a schoolteacher and a mother who was a voracious reader of British Victorian fiction, James came to his investment in "Englishness" through his family's middle-class positioning. For as Reinhard Sanders notes, members of the educated black community (of whom James was one) saw themselves "as part of the tradition of Western European civilization and turned to Europe for its intellectual inspiration and aesthetic standards" (91). "A British intellectual before [he] was ten" (*Beyond a Boundary* 28), James attended Queen's Royal College as a scholarship boy and at the age of fifteen, according to Eugenia Collier, published an essay on cricket and wrote on the theme of the "novel as an instrument of reform" (229), both of which contained, says Collier, "the seeds of his subsequent thought: cricket as a many-sided metaphor; literature as an instrument of change" (229). In 1932 James departed for England, where he stayed for the next six years, traveling to the United States thereafter and then returning to Britain in 1953. During his time in Britain and the United States, James was involved in a variety of progressive and/or revolutionary organizations and causes from the Independent Labour Party to the Pan African movement; during this time as well he produced his immensely prolific writings, including those I look at in this chapter.

Statements expressing his indebtedness to "Western civilization" also point up a distinctive feature of Caribbean cultures, which, according to James, were made up almost entirely of exogenous (read: a large dose of European) elements. In his "Appendix" to *The Black Jacobins,* James notes:

> The West Indies has never been a traditional colonial territory with clearly distinguished economic and political relations between two different cultures. Native culture there was none [*sic*]. The aboriginal Amer-indian civilization had been destroyed. Every succeeding year, therefore, saw the labouring population, slave or free, incorporating into itself more and more of the language, customs, aims, and outlook of its masters. (405)

The figures who emerged from this cultural, historical, and social formation were indeed "Black Europeans." However, as James's remarks on Aimé Césaire reveal, they used their intimate knowledge of and embeddedness within, "Western civilization" to produce devastating critiques of Europe's massive expansion into and exploitation of the lands and peoples it colonized. Using "the master's tools," to adapt a remark from Audre Lorde against its grain, they sought (and often succeeded in their attempts) to "dismantle the master's house."

In this chapter, then, I focus on C. L. R. James and his work in order to delineate a view of resistance to colonialism—its oppressive structures and ideologies—that proceeds from a strategy that inhabits some of those very (metropolitan/colonial) concepts and beliefs through which the colonized peoples' subjugation was secured. James's work, I argue, is marked by an appropriation and simultaneous arm's length—critical and transformative—management of these concepts and beliefs.[3] To make my argument, I focus specifically on his "grand narrative" of the Haitian revolution that James calls "the only successful slave revolt in history," *The Black Jacobins: Toussaint L'Ouverture and the San Domingo Revolution* ("Preface to the First Edition," 2nd ed. ix; hereafter *The Black Jacobins*), his "autosociographical" (Sylvia Wynter's term) study of cricket, *Beyond a Boundary* (hereafter *Beyond*) interspersed with recollections of his boyhood in Tunapuna, his family history and education and subsequent departure for and residence in England, and some essays collected in *Spheres of Existence.* In both *Beyond* and *The Black Jacobins,* as I go on to demonstrate, James takes on the very symbols through which dominant metropolitan/colonial identities and values define themselves—cricket in terms of "Englishness" or the English national character in *Beyond,*[4] and the French Revolution as a symbol of European Enlightenment in *The Black Jacobins*—in order to present his revisionary accounts of the making of a (free) Caribbean people and the San Domingo Revolution respectively.[5] Abiding by James's judgment, itself derived from Marx ("Great men make history, but only such history as it is possible for them to make. Their freedom of achievement is limited by the necessities of their environment" ["Preface," *The Black Jacobins* x]), my examination attends to the set of circumstances that produced James and his work, paying special attention to the context—historical, political, social, and cultural—from which his work emerges, its stated and unstated purposes, assumptions, and audiences. In the penultimate section of this chapter, which is preceded by a section in which I remark on James's self-consciousness about the limits of the kind of resistance he privileges, I also attend, through an analysis of James's explicit and implicit situating of women's concerns and positions, to the limits James is structurally unable to attend to or overcome.

2.

In a recent essay, "Race before Wicket: Cricket, Empire, and the White Rose," Chris Searle has remarked on the integral connections that obtain between English nationalism—England's defining conception of its national identity—and English imperialist culture and cricket: "Cricket being born of Empire was also a vehicle for instituting and instilling Empire" (32). Similarly, "cricket," for James's friend Learie Constantine, "is the most obvious and some would say glaring example of the black man being kept 'in his place'" (qtd. in Searle 35). Certainly, James is aware of this conjuncture between cricket, English identity, and imperialist culture that depended upon a clear, hierarchical separation of "white" from "black" spheres of activity. Page upon page of *Beyond* is devoted to exposing precisely the class and color lines along which the cricket clubs were constituted, reflecting thereby the larger social and political divisions within the Caribbean: the "most part white and wealthy" Queen's Park club; the "exclusively white" and Catholic Shamrock; the "totally black with no social status whatever" Stingo; the "club of the brown-skinned middle class," Maple (to which James belonged); and "the club of the lower-middle class," Shannon, of which Constantine's father was the captain and Constantine himself a member (*Beyond* 55–57). In *Beyond,* James is also consistently critical of the persistent selection of often incompetent white players as captains (because blacks were seen as people who "'cannot be responsible for themselves—they must always have a white man to lead them'" [225]) and the persistent exclusion of competent, often spectacular, black players from West Indian teams participating in international competitions.

However, cricket as a sport explicitly invested in and itself symbolizing the ideals of "a true meritocracy" and "fair play" could be (and was) mobilized by the colonized West Indians in their own interests. In this regard, the sport through which the English sought to define a distinctively English ethos and identity was appropriated and turned against them to embody the colonized peoples' aspirations and self-definitions. Thus, James's account, while exposing the severe and debilitating exclusions West Indian cricket players were subjected to, also demonstrates why cricket (inevitably) became the arena in which black and colored West Indians sought to establish themselves as the equals of their white rulers: "They," says Constantine in the course of an argument with James, "are no better than we" (116).

Initially, of course, this bid for equality expresses itself in the defensive accents of an oppressed peoples: "'You know what I waitin' for?'" asks one of the men of James at the door of the shoemaker's shop, "'When he [Wilton St. Hill, a spectacular Trinidadian batsman from "the lower-middle class" (93)] go to the Lord's and the Oval and make his century there! That's what I want to see'" (97). But this desire for recognition in and by the metropole modulates very quickly into a larger, more subversive claim when West Indian cricketers and the masses who support them are revealed as more "English" than the English themselves. For the English code governing sportsmanship and fair play in cricket is shown as "native" to (because thoroughly internalized by) the West Indians; the masses and the players stringently abide by it:

> The West Indian masses . . . knew the code as it applied to sport, they expected us, the educated, college boys to maintain it. . . . Read the books of Worrell and Walcott, middle class boys of secondary education, and see how native to them is the code. . . . There is a whole generation of us, perhaps two generations who have been formed by it not only in social attitudes but in our most intimate personal lives. (48–49)

But the English, settlers and colonizers alike, who had inscribed the code in the first place as "a mark of differentiation" (49) in the colonies do not; for when they exclude or pass over black or colored players from their teams, or when they choose inept white captains, they violate the code, because they violate its fundamental premise—fair play. And for this they are called to account.

In *Domination and the Arts of Resistance: Hidden Transcripts,* James C. Scott makes the valuable point that "any dominant group is . . . least able to take liberties with those symbols in which they are most heavily invested" (106); taking liberties can, at the very least, invite the charge of "hypocrisy, if not violation of a sacred trust" (102) from those they seek to govern. Bringing attention to such violations, as James does in *Beyond,* functions as a critique that is effective precisely because it grounds itself in the terms selected by the dominant class (in this instance, the English colonizers) to define itself. And it is effective because it renders the colonizers vulnerable by exposing the gap between an ideal conception of an "English" self governed by "fair play" and a sordid reality where such fair play is deliberately ignored in the interests of preserving brute self-interest and power.

However, it is worth emphasizing that James and presumably the West Indian masses virtually never question cricket's ideal of fair play. Rather, in insisting that the colonized West Indians not only live by it but also in many instances surpass the English themselves in adhering to that ideal, James and the West Indian masses incorporate or, better, insinuate themselves within the dominant. This is an important move. Jonathan Dollimore, while specifying different modes of transgression, notes how "at certain historical conjunctures," it "may be more transgressive" to opt for "a strategy of inclusion [/incorporation]." For the subordinate (in this case, the colonized West Indians) "to lay claim to sharing with the dominant" (in this instance, English settlers and colonizers alike) certain values, certain ways of inhabiting the world through cricket "is also to contaminate the dominant's authenticity"—a presumed organic fit between cricket, "Englishness," English colonizers and settlers—"and [thus] to counter its discriminatory function" (51). This "strategy of inclusion [/incorporation]" that Dollimore goes on to label "deviant identifications with the dominant" can, he further notes, "prove as threatening to it (the dominant) as more overtly oppositional identifications" since the "terror of the other may be premised on the terror of the proximate" (52). I would add that the "terror" in this instance is that of the other as a more principled version of the self, governed by and adhering rigorously to an ideal of fair play.

It comes as no surprise, then, that the ideal of fair play got mobilized in the long campaign which led to Frank Worrell's selection, over the white wicket-keeper, F. C. M. Alexander, as captain of the West Indian team's Australian tour. (Not coincidentally, James played a crucial role in the campaign via his paper, *Nation,* in which he made an emphatic and persuasive case for Worrell's captaincy.) Nevertheless, James, in a virtually throwaway comment tucked away unobtrusively in the concluding pages of *Beyond,* both invokes and transforms the English code of fair play; in doing so, he renders it "native":

> I think that for the first, and I hope last, time in reporting cricket I was not fair. But I was determined to rub in the faces of everybody that Frank Worrell . . . was being discriminated against. . . . There was fifty years' knowledge of discrimination behind it and a corresponding anger. . . . According to the code, anger should not intrude into cricket. . . . According to the colonial version of the code, you were to show yourself a 'true sport' by not making a fuss about the most barefaced discrimination because it wasn't cricket. Not me any longer. To that I had said, was saying my final good-bye. (232–233)

In James's remarks above, violation of the code—which is really a refusal to abide by its "colonial version"—by the colonized acquires a quite different valence from violation of the code by the colonizer: fueled by anger stemming from 50 years of knowledge about discrimination, the former's violation paradoxically shows itself to be true to the spirit, if not the letter, of "fair play"; the latter's, however, can claim no such justification, rooted as it most often is in the sordid compromises necessitated by the desire to preserve power.

Furthermore, when in the penultimate sentence of *Beyond,* James makes explicit the fundamental connection between the West Indian players' mastery of cricket and the "case for West Indian self-government" ("Clearing their way with bat and ball, West Indians at that moment had made a public entry into the comity of nations" [252]), he is not only making visible a claim immanent in the work all along, he is also involved in an act of subversion and transformation—producing, as it were, (an)other and "new" knowledge. Homi K. Bhabha, citing "the rule of repeatable materiality, which Foucault describes as the process by which statements from one institution can be transcribed in the discourse of another," notes how "any change in the statement's conditions of use and reinvestment, any alteration in its field of experience or verification, or indeed any difference in the problems to be solved, can lead to the emergence of a new statement: the difference of the same" ("Commitment" 114). By seizing upon a symbol of English (i.e., the colonizer's) national character—cricket—to represent West Indian (i.e., the colonized's) self-definition, James in effect abducts "Englishness" (as defined by cricket) and makes it not the exclusive property of the colonizers but rather the means by which the colonized peoples of the Caribbean set themselves free.[6] Consequently, what we get is the "emergence of a new statement" that destabilizes cricket's ability to represent only the self-definitions of English national character.[7]

3.

A similar process of re-vision and destabilization of the West's "flattering self-portrait" (the phrase is from James Scott 107) operates in an earlier piece of writing by James, *The Black Jacobins.* A historical reconstruction of the Haitian revolt that is at the same time a study of a figure, Toussaint L'Ouverture, who had preoccupied James for several years, *The Black Jacobins* is an attempt to return revolutionary energy and agency to peoples of African descent as a way in part to further the projects of decolonization. In the annotated bibliography to *The Black Jacobins,* under the section entitled "The French Revolution," James comments:

> It is impossible to understand the San Domingo revolution unless it is studied in close relationship with the revolution in France. Fortunately the French historical school of the French Revolution is one of the greatest historical schools of Western civilization. . . . The greatest of them all is Michelet. . . . He has very little to say of the colonial question, but I believe that many pages in Michelet are the best preparation for understanding what actually happened in San Domingo. (383–84)

The explanatory or instructive power for understanding the San Domingo Revolution seems to rest, in the comments above, with the French Revolution and the "French historical school" that deals with it. This logic is all but confirmed by James's choice of the heroic figure of Toussaint L'Ouverture who, in James's reckoning, leads his people to, and therefore is seen as preparing the ground for, the stage immediately preceding full-fledged revolt: A "steward of livestock," L'Ouverture learns about "politics and the military art" from reading "Caesar's commentaries" (91); he first awakens to the idea of the necessary freedom for his people from his reading and rereading of Abbé Raynal's *Philosophical and Political History of the Establishments and Commerce of the Europeans in the Two Indies* (91); his "veneration for France" (247) and his unequivocal belief that "it is under its [that is, the French Republic's] flag that we are truly free and equal" (150) lead his nephew Möise to complain, "it is always in the interests of the metropolis . . . [which are the] interests . . . of the whites" (275) that L'Ouverture speaks.

Yet, in a characteristic move, James notes that although "in the hour of danger, Toussaint, uninstructed as he was, could find the language and accents of Diderot, Rousseau, and Raynal, of Mirabeau, Robespierre, and Danton. . . . in one respect he excelled them all" (198). Whereas these French revolutionaries, because of their class alignments, would "pause . . . hesitate . . . qualify," and compromise their ideals, L'Ouverture—like the West Indian masses and cricketers and like the James of *Beyond,* who took the code of fair play to its logical, uncompromising extreme—"defend[ed] the freedom of blacks without reservation" (198). Similarly, the slogans of the French revolution—liberty and equality—"meant far more," says James, to the black masses of San Domingo "than to any Frenchman" (198). "True" (because uncompromised) revolutionary spirit is thus revealed as a possession of the colonized, not the colonizer. The explanatory power granted to the French revolution is displaced by (now) being anchored instead in the peripheries.[8]

Such displacements are, of course, the integral critical moves most revisionary accounts make. In order, however, to unpack the full significance of these and other displacements in *The Black Jacobins,* we must attend to the purposes and perspectives, stated and unstated assumptions, local and larger contestations that inform James's account of the San Domingo Revolution. Furthermore, to understand the import of his account, we must construct, on the basis of internal and external "evidence," the audience(s) for whom this account is intended.[9]

James has provided three interrelated, explicit statements of purpose for this work: (1) His "Preface" to the second edition (1963) identifies the book as one that was "intended to stimulate the coming emancipation of Africa" (vii), a claim James reiterates in notes included within the work itself (18, 82) and in its concluding pages. (2) The Appendix, entitled "From Toussaint L'Ouverture to Fidel Castro" and added to the second edition, "attempts for the future of West Indies, all of them, what was done for Africa in 1938" ("Preface" vii). From the outset, then, James conceived *The Black Jacobins* as part of, even a stimulant for, revolutionary action and thought that accompanied the projects of decolonization, first in Africa and then in the West Indies. Clearly, an account of revolutionary events from the past is being mobilized in the interests of a present revolutionary aim. (3) His "Foreword" to the 1980 edition of *The Black Jacobins* published by Allison and Busby in Britain, while undoubtedly informed by hindsight and several years of involvement in the struggles for decolonization, provides a more capacious but also more pointed definition of his earlier statements:

> I was tired of reading and hearing about Africans being persecuted and oppressed in Africa, in the Middle Passage, in the U.S.A. and all over the Caribbean. I made up my mind that I would write a book in which Africans or people of African descent instead of constantly being the object of other people's exploitation and ferocity would themselves be taking action on a grand scale and shaping other people to their needs. (v)

Add to all these statements James's clear sense of a *lack*—both in terms of a suppressed or excluded history, and when it is not excluded, then the absence of "a comprehensive and well-balanced treatment," stemming from "dishonest" colonialist "misrepresentation," "unscrupulousness," and the "white bias of the time" (*The Black Jacobins* 381, 382, 380, 386). From this lack arises James's consciousness of the need to redress the balance, contest falsehoods, set the record straight. All this fuels and is itself fueled by a perspective that, although interested in disclosing "close parallels, hitherto unsuspected" between the revolutions in France and San Domingo, is nevertheless primarily interested in disclosing the revolutionary capacities of "Africans and people of African descent" to forge their own freedom. Its sights are set elsewhere.

Who needs to be informed of this lack or absence? In other words, who, by James's reckoning, are his readers? As I demonstrate below, his readers are primarily white, but some are from the Caribbean diaspora based in the metropole, and this factor, in addition to those mentioned above, shaped the narrative he wrote—the direction his arguments took, the materials, characters, events that he selected for attention and emphases.

What happens when readers who have been brought up on a steady diet of accounts produced by and from the perspective of the colonizers about the colonized are suddenly confronted with an account whose declared interests reside with the colonized and that is thus involved in producing (an)other "truth"? Readers who do not summarily dismiss this alternative account will at the very least be called upon to re-adjust, and ultimately perhaps to re-align, their perspectives. Such re-adjustments and, finally, re-alignments are not easy to provoke or achieve. James is fully conscious of the difficulty of his task, which explains to some extent what I call the self-consciously rhetorical aspects of his work. Since his readers had to be persuaded of an alternative "truth," the standard or dominant account's "truth" had to be dislodged, its interests named, its seamless and authoritative surface disrupted. This rhetorical exigency, I believe, explains (in addition to James's desire to set the record straight) his overt and often polemical contestation of standard colonial sources and representations about "Africans and people of African descent."

Thus, for example, James opens his narrative by disposing of the myth that "the African slave in America was happier than in his own African civilization" (7); he describes the inordinate suffering of the slaves during the Middle Passage (8–9) as he also describes the hypocrisy of the colonists and settlers and of the British (71, 227); he delineates in detail the cruelty and barbarism of the white planters (103); and he asserts the slaves' essential humanity (11–12). Such assertions seem self-evident today; but James uses them as a way of initiating what Robert Hill thought of as "the destruction of accepted scholarship in regard to the Abolition question in England" (27). While revealing the sordid history of self-interest that influenced

British calls for abolition (50–53), James remarks on how "a venal race of scholars, profiteering panders to national vanity, have conspired to obscure the truth about abolition" (51).

While James is uncompromising in his criticisms of the brutal self-interest that underwrote the self-justifying scholarship on colonial rule and thus runs the risk of offending some (or all) of his metropolitan readers, he nevertheless accommodates some of their needs. This explains in part (and only in part) his attempt to anchor his account of the San Domingo Revolution in the discourses and values of the Enlightenment that informed the French Revolution. For by doing so, he is invoking something they are likely to be familiar with and would presumably value. In this regard it is equally important to note that the rhetorical exigencies of his task coincided with his own political development. For not only was his primary audience situated in the metropole; the precise contours of his argument and vision were also forged in the metropole. He says as much in *Beyond* when he points to his increased politicization, via his increased contact while living in Nelson, Lancashire, with an already radicalized Constantine, and to his growing knowledge about the West Indies: "We were educating thousands, including ourselves" (121). What was the content and nature of this education? Recalling how "Constantine by his cricket . . . had created an enormous interest in the West Indies and in West Indians" in Britain, James observes how "most people were hazy about both the islands and the people" (120). However, what could easily have been a task that involved simply providing information became, in the colonial context, an overtly political burden: to demonstrate that "we [James, Constantine, and other West Indians] were as human as the rest of them [the English]" (121). It was at this point, says James, that

> West Indian history . . . began to assume a new importance. Stuck away in the back of my head for years was the project of writing a biography of Toussaint Louverture [*sic*]—the leader of the revolt of the slaves in the French colony of San Domingo. This revolt and the successful establishment of the state of Haiti is the most outstanding event in the history of the West Indies. (122)

Though one cannot and should not diminish the significance of this history for "Africans or people of African descent" (or, for that matter, any other colonized or oppressed groups),[10] its initial significance seems to reside precisely in the task of demonstrating to its metropolitan readers, ill-informed or simply ignorant about the West Indies, that West Indians were as human as the English, could be agents of their own freedom, and indeed could write themselves into, by becoming subjects of, their own history. To persuade them James had to appeal to their values, their belief systems, and their "grand narratives."

Here it may be helpful to specify that in deciding to center the project of anti-colonialism in the metropole, James is not the exception but the rule. People like Ras Makonnen, George Padmore, and Eric Williams from the British-ruled Caribbean, and Aimé Cesaire, René Maran, and Frantz Fanon from the French-ruled Caribbean followed a similar path. Furthermore, as the case of Présence Africaine, the publishing house, and *Présence Africaine,* the journal, demonstrate, at this historical juncture (the late 1940s onward; the journal was founded by Alioune Diop, "a French educated Black African" [Mudimbe xvii] in Paris in 1947) the metropole was perceived as the definitive site for launching the case against colonial rule. Christopher Miller remarks on the "enabling" ambiguity of this paradoxical move when he observes:

> Created at the apogee of French colonial power, Présence Africaine, the publishing house, and *Présence Africaine,* the journal, were conceived within the terms and conditions of hegemony, but as a first step toward the dismantling of colonial power from within. The Présence Africaine was founded and remains in Paris, the center of the colonial empire, absent from Africa whose presence it began to pursue, describe, and advocate. Nothing could be done about France without France. . . . Without Paris, then, without the absence of Africa, there would be no Présence Africaine. ("Alioune Diop" 427)

V. Y. Mudimbe emphasizes the propitious historical conjuncture—"the establishment of *Présence Africaine* coincides with a new will to truth in European consciousness"—and ideologically compatible "traveling companions" among some of the French intellectuals, to assert *Présence Africaine*'s success in making "highly visible a marginal discourse and its dreams" by "situat[ing] itself forcefully on a level of political and ideological struggle in Paris" (443).

With some variations, what Miller and Mudimbe observe about the context(s) from which the projects of *Présence Africaine* emerged could be observed as well of the projects undertaken by figures from the British Caribbean, including James. One additional factor needs to be specified as to why the projects for dismantling colonialism were seen by many Caribbean radicals as projects that had to be necessarily situated in the metropole: "To start a fight [in the colonies] at the time couldn't be done," says James in the course of an interview with *Kas Kas:* "Today they [that is, the radicals] are going to remain at home" (39). As Robinson puts it, "the center of the British Empire was the most significant platform . . . [because]

colonial radicals could depend upon the British tradition of free speech and a free press in their attack upon the Empire" (*Black Marxism* 372). However, these freedoms, simply assumed because they were easily accessible in Britain, were not accessible in the colonies. James's contemporary and friend, Ras Makonnen, for example, was conscious of this disparity:

> What was it like to be a Black man in the Britain of 1930s? . . . Imagine what it meant for us to go to Hyde Park to speak to a race of people who were considered our masters, and tell them right out what we felt about their empire and about them . . . write any tract we wanted . . . all this when you knew very well that back in the colonies even to say 'God is love' might get the authorities after you! (Qtd. in Robinson, *Black Marxism* 372–373)

Thus Makonnen correctly points out:

> When you look at the results of those Africans who had been to England, you wouldn't be far wrong in saying that England had been the executioner of its own colonial empire . . . allow[ing] these blacks to feel the contrast between freedom in the metropolis and slavery in the colonies. (Qtd. in Robinson, *Black Marxism* 373)

4.

James's critique of empire, launched from and within the center of that empire, its discourse and the terms of its contestation shaped by the West's self-definitions, raises an important question: Are we dealing here with something akin to Michel Foucault's conception of resistance as something inscribed within the operations of power itself—resistance, that is, as a creation and effect of, and therefore dependent on, power?[11] Of course such a conception of resistance tends to rob resistance, as critics of Foucault point out, of its oppositional force. Thus Said quite rightly complains that "resistance cannot equally be an adversarial alternative to power and a dependent function of it, except in some metaphysical, and ultimately trivial sense" ("Travelling Theories" 66). James's critical project(s), I believe, can be better defined by a somewhat different conception of resistance than Foucault's. Raymond Williams, for example, while granting that "dominant culture . . . at once produces and limits its own forms of counter-culture," also remarks on the possibility of an oppositional or resistant view and practice that signify a genuine break from dominant culture:

> It would be wrong to overlook the importance of works and ideas which, while clearly affected by hegemonic limits and pressures, are at least in part significant breaks beyond them . . . which again may in part be neutralized,

> reduced, or incorporated, but which in their most active elements nevertheless come through as independent and original. (*Marxism and Literature* 114)

In a sense, the various displacements and disruptions in James's work of official colonial accounts and of European (primarily British and French) self-definitions that I have looked at so far signify precisely such "significant breaks." Certainly such breaks are a function of James's perspective, his attention fixed steadily on the colonized Africans and West Indians and on their legitimate grievances and bid for freedom. But they are also a function of *how* James locates himself (and indeed sees himself as located) with respect to the metropole. Here something James observes about diasporic populations from Africa and the Caribbean in Britain is instructive:

> They [that is, diasporic peoples] will be intimately related to the British people, but they cannot be fully part of the English environment because they are black. Now that is not a negative statement. . . . Those people who are *in* Western civilization, who have grown up *in* it, but [have been] made to feel and themselves feeling that they are *outside,* have a unique insight into their society. (Qtd. in Mercer, "Diaspora Culture" 56)

"*In* the West, but not [necessarily] *of* it," that is, enclosed but not included in the host,[12] diasporic peoples (specifically, diasporic peoples of color who are excluded from the possibility of comprehensive assimilation within the host culture by virtue of their alleged racialized status and identity) possess a "unique insight" that is a product of their critical distance from exactly the structures of knowledge and ways of being they inhabit so intimately.[13] They are like those "liminal groups" identified by Asmarom Legesse, whom Sylvia Wynter quotes, who are most able to "free us" from the dominant regimes of "truth" because "it is they who experience the 'injustice inherent'" in these regimes ("The Ceremony Must Be Found" 38). Thus James, as someone situated within *and* outside the metropole, can simultaneously assert that "the origins of my work and my thoughts are to be found in Western European thought" ("Discovering" 237) *and* be "a declared enemy of British imperialism" (*Beyond* 54); or, as he says elsewhere, "I denounce European colonial scholarship. But I respect the learning and profound discoveries of Western civilization" ("The Making" 179). Thus, too, James says of L'Ouverture:

> It is Toussaint's supreme merit that while he saw European civilization as a valuable and necessary thing and strove to lay its foundation among his people, he never had the illusion that it conferred any moral superiority. He knew French, British, and Spanish imperialists for the insatiable gangsters that they were. (*The Black Jacobins* 271)

What is remarkable about James is that though he himself (and the character he selects as the "hero" of *The Black Jacobins*) represents a figure whose relationship with "Western civilization" is complex, fueled by the paradoxical drive to both engage with and distance himself from, he is nevertheless able to present (sympathetically) figures who function virtually *outside* of the logic of "Western civilization" either because they refuse or are not called upon to engage with the "contradictions of domination, the dialectic of imperialism" (Robinson, "Amilcar Cabral" 40). Through them James enables his readers (and himself, for that matter) to examine his own formation and that of others like him, whose resistance took the form it did because it was produced by a particular working out of the colonial encounter. For example, in a work like *Beyond,* which delineates both the boundaries (that is, limits) within which the West Indians achieved their freedom and the transcendence of those limits (which *beyond* surely implies), it is appropriate that James begins his narrative with a figure like Matthew Bondsman who exists *outside* the pale of respectable middle-class society to which James and his family aspire: A "young man" of "medium height," Matthew Bondsman is described in the opening pages as an "awful character . . . generally dirty . . . would not work . . . his language was violent . . . his voice . . . loud" and he would "without shame walk up the main street barefooted" (14). But what is evident in James's presentation is that Matthew is an outsider by choice: he refuses to abide by a bourgeois work ethic, by bourgeois values of cleanliness and proper speech and behavior. Though a fine batsman (for which ability James labels him "*genus Brittanicus*" [14]), Matthew Bondsman refuses to dedicate himself to that task as well: "He would not practice regularly, he would not pay his subscription to the club" (14). By refusing to nurture his natural talent, of course, Bondsman is refusing even a minimal incorporation into hegemonic, colonial structures that James himself settles for. As such, then, Bondsman is beyond (that is, outside) the boundaries of the ideas of resistance that the book maps. But by allowing Bondsman and all that he stands for to initiate his narrative, James scrupulously inscribes and attends to (an)other and different mode of resistance.[14]

In *The Black Jacobins,* the character who is explicitly contrasted with Toussaint L'Ouverture, Dessalines, serves a similar outsider-function:

> Yet Toussaint's error sprang from the very qualities that made him what he was. . . . If Dessalines could see so clearly and simply, it was because the ties that bound this uneducated soldier to French civilization were of the slenderest. He saw what was under his nose so well because he saw no further. Toussaint's failure was the failure of enlightenment, not of darkness. (288)

In this remarkable passage, where James both concedes L'Ouverture's blind spot and rationalizes it, he nevertheless scrutinizes "the ambivalent pride of place [he has] presumed for the Westernized ideologue" and thereby scrutinizes his own formation as well (Robinson, *Black Marxism* 385). He does so by invoking the "virtues" of Dessalines, which spring precisely from his being outside the influence of French civilization. (It is not exactly clear, in this passage or elsewhere in the work, whether this is for Dessalines a matter of choice.) Thus the final, revolutionary break from French colonizers is made by Dessalines, not L'Ouverture, even though L'Ouverture had been instrumental in revolutionizing the masses and in preparing the ground for that break to take place. For Robinson, L'Ouverture's "failure" represents James's recognition of the "boundaries beyond which the petit bourgeoisie could not be trusted." For that reason, Robinson adds, James "was to insist so often that the revolutionary masses must preserve for themselves the direction of the revolutionary movement, never deferring to professional revolutionists, parties, or the intelligentsia" (*Black Marxism* 386).

5.

Despite James's ability to represent, via Bondsman and Dessalines, positions and/or characters that function outside the boundaries (that is, limits) of what I have described as the prevailing method and cultural/historical formation underpinning his vision of resistance against colonialism, there are nevertheless other boundaries beyond which his "grand narratives" of resistance and emancipation do not, perhaps cannot, go. These limits are best described through an analysis of a conception of gender[15] that haunts both *The Black Jacobins* and *Beyond,* implicitly in the case of the former and explicitly in that of the latter.

That James's rewriting of "lily-white" dominant history in *The Black Jacobins* remains male centered (and male determined) is perhaps to state the obvious given his exclusive focus on heroic male figures and their actions leading to the freedom of the slaves in San Domingo. Such an emphasis might well be a product of his investment in appropriating, albeit for his own ends, the values and method that underpin dominant history. And dominant (or, for that matter, all traditional) history locates itself and is invested in the "public" realm, whereas "the history of women and gender relations," Rosalind O'Hanlon reminds, "are assumed to occupy an area of the historical field rather distant from these larger political processes" (94). "Assigned to the more limited [because presumably "private"?] domains of family organization, reproduction and sexuality," O'Hanlon adds, "and singled out as objects of specialist interest at best supplementary to more established frameworks for political and social investigation," women and gender relations can be, and often are, devalued, when not altogether ignored. Yet Stella Dadzie in her study of slavery and resistance in Jamaica notes that "slavery and slave trade, providing as they did a crude levelling off of sexual distinctions, ensured that the African woman, 'shared every inch of the man's spiritual and physical odyssey'" (21). She shared something else as well: her resistance against the severe debilitations of slavery and a brutal, and brutalizing, plantocratic culture was not only confined to what O'Hanlon calls the arena of "reproduction and sexuality" but also made itself evident through what are usually characterized as "male" acts of violence and turbulence—"malingering, insubordination, strikes and physical assaults [on plantation owners, but especially on the overseers]" (Dadzie 26–31; 24). And if the African woman's active participation in insurrections and outright revolt was not as extensive as that of her male counterparts, she often "acted as go-between liaising with rebel slaves" (Dadzie 32). While it would be misleading simply to substitute San Domingo for Jamaica and read Dadzie's remarks about women's participation in the "public" (presumably "male") realm in the latter colony as instances of the same in the former, it would not be misleading to speculate about that which remains completely outside the bounds (because deliberately suppressed?) of James's narrative about "the only successful slave revolt in history." Were women, in fact, absent and/or nonparticipants in this particular revolt, or is their absence a narrative choice, an example of an historical method whereby women are "caught up by default in [an account] of confrontations from which only the men could emerge as heroes"? (Dadzie 21).[16]

The division between "public" and "private" spheres of activity so integral to the definition and construction of the (different) roles assigned to men and women, and the view that history is exclusively concerned with the larger political and social processes that characterize the "public" realm explicitly and implicitly underwrite the narrative of *Beyond* as well; but they function here in ways that enable its readers to deploy gender as a category internal to its analysis. The "public" and the "private" realms in *Beyond,* though often explicitly presented as differentiated and mutually exclusive realms for the definitions of male and female roles, do not ultimately remain hermetically sealed within their own boundaries; they are instead porous categories, not altogether impervious to traffic between and within one another, giving rise thereby to a conception of gender fraught with all sorts of ambiguities, contradictions, and displacements. But these contradictions remain buried in a narrative invested in a primarily male-centered view of what constitutes the "public" and the "private," with the "public" being the realm within which definitions of resistance are elaborated. These contradictions need to be teased out and exposed for us to grasp the full force of James's overt and covert evasions.

James begins by locating the contributions and participation of women (his grandmother, aunts, mother, and sister) squarely within the realm of his private history and development. Conversely, the "public" realm is defined through cricket—cricket not as mere sport, but cricket-as-social-relations and as instrument for West Indian self-determination and elaboration of a West Indian "national" identity. Insofar as cricket is (implicitly) specified as a "male" endeavor, and insofar as women, specifically James's "grandmother and aunts," are "not in any way supporters or followers of the game" (15), they seem, at first glance, to be closed off from participation in cricket-as-social-relations. These divisions between the "private" and the "public," moreover, are reinforced when James notes how his aunt Judith's son,

> though no active sportsman himself, once a year invited his friends from everywhere to Tunapuna where they played a festive cricket match. This, however, was merely a preliminary to a great spread which Aunt Judith always prepared. (21)

In this passage, though cricket is assigned the second place, its attractions coming second to those of Aunt Judith's "great spread," Aunt Judith is nevertheless consigned to the domestic (and female) task of providing the meal, her access to cricket and therefore, the "public" realm mediated by her role as provider and as the mother of her son who, despite not being an "active sportsman," can nevertheless play in "a festive cricket match." In this regard, it's also worth noting a further instance of such differentiation based on access (via cricket) to the "public" realm. While Matthew Bondsman, despite being an outsider with respect to the conception of resistance James elaborates, can be inserted into James's narrative of emancipation, it seems the women cannot. Why? Because "Matthew had one saving grace—he could bat" (14).

But if women are excluded from the "public" realm at the literal level because they neither play nor support nor follow cricket, they are not excluded at the figurative level, because their very beings, self-definitions, and activities compel connections with the ways in which James defines cricket:

> What was it that so linked my Aunt Judith with cricket as I, a colonial, experienced it? The answer is in one word: Puritanism; more specifically, restraint, and restraint in a personal sense. (47)

This restraint, James emphasizes, learned on the "cricket field," and via "*The Captain* and the *Boy's Own Paper,*" the pervasive impact of "the university men who taught me," and, of course, Aunt Judith and his mother, "was in essence the same code—and I was learning it very early from *Vanity Fair*" (47). In this passage, where James not only sketches in the significance of women to his own development but also links them with the code of personal restraint that characterizes cricket, he identifies one other affiliation in this chain of comparisons—Thackeray's *Vanity Fair.* But what James gives with one hand, he takes away with the other. Thus, though it is his mother's "old copy with a red cover" that introduces him to *Vanity Fair,* it is his father he credits, in fundamentally gendered terms, with helping him cultivate "discrimination" in reading; his mother's reading habits, he says, are "indiscriminate" (26).[17]

Given such a contradictory assessment of women, who are and are not participants in the "public" sphere within which James casts his narrative of resistance, it is not altogether surprising that the "public transcript" documenting the fundamental and nearly exclusive role of men in West Indians' emancipation from colonial rule is at odds with the "hidden [or barely visible] transcript"[18] that insinuates women into the same narrative of resistance, albeit by surrounding their ability to participate with all sorts of qualifications and disclaimers. Scrupulous attention to the ways in which James's narrative negotiates this contradiction enables us to read against the grain of his preferred definitions about his work. More importantly, scrupulous attention to the (inevitable) contradictions that surround his representations of women and their roles in society brings to the surface (an)other occluded but nevertheless potential narrative that hints at, though it does not unequivocally endorse, women's access to and participation in the narrative of cricket-as-social-relations, which is also the narrative within which the West Indian resistance to colonial rule is mapped out.

6.

If James's non- (or at best contradiction-ridden) inclusion of women in his account signifies the limits of the narrative paradigms he selects, the world view and historical moment he inhabits, it does not therefore vitiate or cancel out his contemporary significance with respect to our understanding and critiquing of colonial/Western/dominant discourses and power in what, with some hesitation, I concede is a "post"-colonial era.

Walter Rodney notes how James often focuses on little-known or virtually unacknowledged resistances and social movements, seeing them as "part of the sociopolitical protest engendered by the presence of the Europeans and systems of colonialism" (5). Through such a focus, James not only calls into question the colonizers' claims about the impermeability of their power, but more significantly he restores subjectivity to the colonized, making them agents of their own violated but recuperable histories.

By doing so, James also brings our attention to what would otherwise remain excluded, suppressed, or neglected histories and perspectives. Finally, by reflecting on why these exclusions occur in the first place—plain dishonesty, or unscrupulousness, or uncritical faith in officially sanctioned accounts and memory—he interrogates and undermines the power that underwrites dominant accounts and self-definitions. All these critical moves should be of significance to scholars of colonial and postcolonial discourses.

If there is a persistent claim that recurs in several poststructuralist analyses of oppositional visions and politics, it is this: even though "the drive behind [the oppositional] mode seems to be the desire to find some stance outside the prevailing system of representation [values and ideology], from which to impeach and castigate [them], from which to overturn and transform. . . . such an outside point is not so much found ready-made 'out there' as it has to be painfully constructed from within" (Maharaj 15). Constructing an oppositional vision and politics often entails what Kobena Mercer describes as a double-pronged strategy: critically appropriating "elements from the master-codes of dominant culture . . . disarticulating given signs and rearticulating their symbolic meaning otherwise" ("Diaspora Culture," 56). James's oppositional politics, born of the colonial encounter, at once intimately engaged with and appreciative of "Western civilization," yet fundamentally critical of its exclusions and brutality, pointedly illustrates and thereby provides the tools for achieving this and other similar ends.[19]

Finally, the fundamental contemporary significance of James—of the projects he undertook, the positions he mapped out, the allegiances he paid homage to—is this: at the current historical moment, when a politics of identity (not restricted, I would argue, to essentialist stances peculiar to oppressed groups) is mobilized crudely and reductively to set up figurative and ultimately coercive boundaries between the "West" and "the-rest-that-is-not-the-West," it is bracing to encounter someone like James, to whom such boundaries would have been unhelpful. Rather, he asks his readers "to see the enemy in all its amplitude" ("Two Young" 112), which entails rejecting, or at the very least interrogating, any reductive claims; hence his refusal, for instance, to reduce all Western culture to its imperialism. Instead, he demonstrates both what it means to be a partisan of his own peoples (and therefore what it means for someone to come from a colonized part of the world) and what it means to belong to, indeed possess, other allegiances. And while doing so, he makes us see that there is no necessarily innate contradiction between the two.

Chapter 2

Re-Playing the Indian Subcontinent: Salman Rushdie's Method(s) of Critique

Between colonialism and its genealogical offspring, there is, thus, a holding and a crossing-over. Many of the post-colonial writers bear their past within them—as scars of humiliating wounds, as instigation for different practices, as potentially revised visions of the past tending towards a future, as urgently re-interpretable and re-deployable experiences in which the formerly silent native speaks and acts on territory taken back from the colonist.

—Edward W. Said

Yuri Zynianov . . . points out that "stylisation and parody are closely linked to each other: Both live a double life. Beyond the work, there is a second level, stylised or parodised. When stylisation is strongly marked it becomes parody." . . . He defines parody as "the mechanisation of a particular procedure, a mechanism which, of course, will only be noticeable where the procedure to which it applies is known. Hence, if parody's style is not sufficiently familiar to the audience, they will be unaware that something is being parodied."

—Paul Willemen

1.

In his review article on Salman Rushdie, Anthony Barnett's criticism of *Midnight's Children* (hereafter *MC*) indicates what seems to me to be a simultaneously astute and mistaken understanding of Rushdie's representation of the Indian subcontinent. Prefacing his comments with a passage from Margaret Drabble's report on her recent visit to India, a report he dubs "the quintessence of tourism," Barnett goes on to remark on "all the romanticism" *MC* "evoked, despite itself":

> . . . the teeming multitudes of India . . . their oddness, their magical strangeness, their—now we should perhaps use the word only within the museum frame of inverted commas—their "humanity." . . . There is little there, one can see it from one's window [alluding to Drabble's "simply to gaze from one's window was excitement enough"]. . . . This attitude has always been around, only now it is fashionable. . . . What I am criticizing is the "appreciation" of the coloured multitudes as a spectacle without risk. ("Salman Rushdie" 95)

Buried in these comments about Rushdie's shortcomings in *MC* are certain assumptions about what constitutes apt (because "authentic") descriptions of India, especially when they emanate from a writer from a formerly colonized country. Such a writer must not only be "defiant" of the (colonizing) West, but his defiance must also be expressed through an implacable opposition to the stereotypes about the colonized the West mobilizes in its own representations. How can Rushdie, Barnett seems to be complaining, recycle these stereotypes? Furthermore, how can Rushdie (an insider, because "native" to India) view India panoptically, delivering its "coloured multitudes as spectacle," which, after all, is the viewpoint adopted in the writings of the former colonizers and their contemporary avatars, British tourists like Drabble, who are also outsiders to the countries/cultures they visit and describe? Barnett's remarks assume that there is an inside and an outside position one occupies with respect to cultural representations, and that these positions are mutually exclusive, especially when they pertain to representations of the (formerly) colonized world. And his remarks assume that an inside position is automatically oppositional to, as an outside position is automatically collusive with, colonial/Western representations.

The implicit categories that organize Barnett's criticism of *MC*—in which the dichotomy that sets inside against outside also evokes other oppositions between authentic and inauthentic, native and foreign—resonate with (and within) other critical or skeptical readings of the dissident potential or progressive political freight of Rushdie's work. In virtually each of the critical/skeptical readings, Rushdie's ability to elaborate a genuinely resistant or emancipatory alternative to hegemonic, Western descriptions of the subcontinent is rendered suspect by being tied to what is construed as Rushdie's problematic location as an immigrant from the subcontinent now residing in and a citizen of Britain, the subcontinent's former colonizer. Born to an affluent, Muslim family in Bombay in June, 1947, "only two months," Rushdie is fond of saying, before the British departed from India, educated, first, at Cathedral School in Bombay, a classic (neo-)colonial enclave that prepared him for where he was headed next, the exclusive British public school, Rugby, and England, Rushdie has spent most of his life, certainly his entire writing career, in Britain. However, location, as the critical literature on the subject makes clear, is not a simple matter.[1] Tied, in Rushdie's case, to his work's embeddedness within Western epistemologies and "regimes of signification" (Paul Willemen's phrase), his location is viewed as signaling most often Rushdie's lack of allegiance to specifically "native"—that is, subcontinental—agendas and investments. Rushdie's problematic location, then, serves as the grounds on which his right to represent the subcontinent is seen to rest.[2] In Rushdie's case, of course, this criticism/skepticism is exacerbated by the fact that in *MC* and *Shame,* he is writing not only broad, sweeping narratives of the two nations, India and Pakistan, but also narratives that are bitter critiques, acerbic denunciations of what these nations have become, from his physical location outside of rather than from within the subcontinent.

Contrary to what his critics aver, Rushdie has always insisted on the writer's adversarial relationship with dominant ideologies, and he has described his own work in particular as proceeding from the desire to represent alternative, emancipatory perspectives—perspectives, that is, that are excluded, suppressed, or marginalized by dominant representations.[3] Furthermore, while conceding that his own position is that of an "insider and outsider in society," Rushdie notes that in *MC* and *Shame* he "absolutely did not want to write an outsider novel" ("A Dangerous Art Form" 3–4).[4] How are we to square this obvious disjunction between Rushdie's views of his own work with those of his critics?

One way to proceed would be to mine the insider-outsider position not for its negative or ambivalent freight alone, but for the "productive unease" (Gayatri Chakravorty Spivak's phrase) it can generate. In the preceding chapter, for example, I allude to the "unique insight into [British] society" that C. L. R. James ascribes to diasporic populations from Africa and the Caribbean within Britain. This "insight" he sees as a product of their being both inside and outside: "*in* the West, but not [necessarily] *of* it." In Rushdie's work, comments like these acquire a more expansive focus. Restricted no longer to the dialectic of belonging/non-belonging, identity/non-identity the migrant feels (and mobilizes for critical insight) with respect to her or his adopted "home" (Britain), the dialectic pertains as well to the "home" she or he has left behind (India, Pakistan). While Rushdie, in an essay celebrating "the migrant sensibility," emphasizes the virtues of being positioned outside ("To see things plainly, you have to cross a frontier" ["The Location of Brazil" 53]), his sense of himself as someone whose identity is mediated by and itself mediates both Britain and the subcontinent has much in common with the position James ascribes to diasporic populations.[5] In both James and Rushdie, in other words, critique and contestation are viewed as proceeding from within the structures of knowledge forwarded by hegemonic formations, but by those people who perceive themselves (and/or are perceived) as not belonging (or not belonging entirely) to and who therefore *position* themselves as also existing outside these formations. However, an important disclaimer is in order. Whereas, for James, the "unique insight" he ascribes to diasporic populations is merely one of the positions he mines in order to elaborate his oppositional politics, in Rushdie's work his position as a migrant blessed with "unique [critical] insight" is made into "a singular repository of experience and resistance" (Krishnaswamy 126).

Intimately tied to this awareness of critical insight as a product of location that is at once within and outside hegemonic formations is a discursive practice of resistance that appropriates concepts and/or representations of what Willemen calls "dominant regimes of signification" to subvert belief in their truth claims. In this chapter, I analyze Rushdie's appropriation of the dominant through two interrelated strategies. The first I characterize as a strategy of re-play, in which what is appropriated is literally repeated and re-staged for critical scrutiny. (One of the risks in, or limits of, using this strategy is, as Barnett's criticism demonstrates, that if the parody of the "original" is not recognized as parody, or if the "original" is not recognized at all, then the critical insight re-play brings into play is rendered meaningless.) The second strategy I elaborate entails unearthing exclusions (including suppressed accounts and/or perspectives) in dominant accounts/formations to: (a) reveal lacunae that in turn undermine the dominant's invincibility; (b) insinuate a different, sometimes oppositional perspective, because the dominant is re-played from the point of view of the excluded—here we move into the arena of what could be called alternative or counter representation.[6] Insofar as these strategies are mapped in works whose ambition is to (re)write the histories of two nations, they engage with (and critique) as well the methods of and faith in objective facts that characterize traditional historiography.

2.

Barnett is on the mark when he registers a sense of déjà vu while reading *MC.* He has seen Rushdie's India before, most recently, as he suggests, in the report of a British writer, Margaret Drabble: "the teeming multitudes of India . . . their oddness, their magical strangeness" have indeed been staple features in the work of colonialist/Western (and deracinated subcontinental?) writers. Contra Barnett, however, I would argue that this repetition of familiar representations in Rushdie's work should not be read as a simple or uncritical repetition (as I think Barnett does); rather, it should be viewed as a deliberate strategy of re-play that in fact inhabits dominant representations in order to deconstruct their claims to absolute truth or reality.[7]

At the most basic and uncomplicated level, re-play entails re-viewing, or going over again, with a view to looking carefully for something that may have been overlooked at first sight. It is a re-staging that, in terms of mimetic descriptions, suggests that the "original," or prior representation is itself something constructed, that what has been staged as immediately "real" is itself a mediated presentation, that is, a *re*-presentation. Re-play's most obvious connection is with the cinematographic, a connection that Rushdie invokes in a passage linking the genesis of *MC* with his (re-)visit to Bombay that captures some of the resonances I wish to evoke when I use this term. As a procedure that signals an account or a representation that comes *after* and follows an "original" or a prior representation, my definition of re-play will, in the course of this chapter, acquire a sense comparable with what Homi Bhabha has defined as "mimicry" and as a "supplement" or "supplementary." In a colonial context, "mimicry" for Bhabha registers the ambivalence of colonial discourse, the slippages and displacements this ambivalence enables. "Mimicry," never simply a copy of a colonial "original," is the process by which the colonized subject (mis)appropriates the terms of colonial discourse, thereby interrogating and resisting their authorizing presence. (On "mimicry," see especially his "Of Mimicry and Man" and "Signs Taken for Wonders.") Similarly, with respect to the "supplement," Bhabha notes how

> being "after" the original, or "in addition to it," gives it the advantage of introducing a sense of "secondariness" or belatedness into the structure of the original. The supplementary strategy suggests that adding "to" need not "add up" but may disturb the calculation. . . . Insinuating itself into the terms of reference of the dominant discourse, the supplementary strategy antagonizes the power to generalize, to produce sociological solidity. ("DissemiNation" 305–306)

I use "re-play" rather than "mimicry," or "supplement," or "supplementary" to highlight the significance of the visual and cinematic in Rushdie and to identify specific representational techniques employed by him.

In a passage in his essay "The Indian Writer in England," in which Rushdie alludes to the genesis of *MC,* he compares his actual sight, while re-visiting Bombay, of the house in which he was born with a photograph in black and white of the same house in his study in London. Rushdie notes how his "memory feeding on such [black and white] images as this, had begun to see [his] childhood in the same way, mono-chromatically. The colours of my history had seeped out of my mind's eye." On (re-)visiting this house, however, his "other two eyes were assaulted by colours." "It is probably not too romantic to say," he adds, "that that was when *Midnight's Children* was really born; when I realized how much I wanted to restore my past to myself . . . whole, in Cinemascope and glorious Technicolour" (75).

Re-visiting Bombay allows Rushdie to re-view/revise a past representation encapsulated in an old photograph of the house in which he was born, and which, during his years of absence from the subcontinent, served as an important link and had perhaps become identified with India itself. Its black and white presentation suggesting plain, uncolored truth and documentary "reality," this photograph, in the absence of other evidence and through its sheer availability, had acquired the effect and force of the "real." However, what emerges via second sight—which, in this instance, is his sight of the actual house—and the introduction of color is the impoverishment and diminution of the "original" or prior representation of the house in the photograph. Rushdie's impression of the photograph's inadequacies is based on his awareness of what it excludes and cannot capture—a recognition that simultaneously establishes the "original" as only reproduction, an artifice that reflects the limits of its composition. However, second sight—his sighting of the actual house—for all its immediacy (he describes the "vividness of the red tiles, the yellow-edged green cactus leaves, the brilliance of the bougainvillea creeper" [75]) cannot be taken as "real" either. It is a product of his sensory and emotional deprivation and is "coloured" by his desire to reclaim the past that is Bombay/India. Unlike the understated, monochromatic photograph, it is hyper-real—more and, at the same time, less "real." Rushdie does not conceal this fact; indeed, he calls attention to it when he invokes Cinemascope and Technicolor—techniques to simulate the real through heightened effects—as analogous procedures to those by which he will carry out his project of (historical) reclamation. (This re-visit, including especially his actual sight of his house, incites in him the desire to "reclaim" a "city and a history," which he proceeds to do by writing *MC* ["Indian Writer in England" 76].) Further, by evoking technical innovations, those involving *techne,* or art, he underscores that his will be an artful re-creation of objects forwarded to leave a particular impression of reality (through expanding one's field of vision and frame of reference—Cinemascope—or through "colouring" representations—Technicolor—or through setting in motion a group of stills to effect a moving picture).[8]

Beyond the formal features of re-play and their ideological effects of the constructing and constructed nature of all representations that this personalized, anecdotal account of *MC*'s genesis elaborates, this account is followed as well by another use of re-play via a specific intertextual reference that has escaped all commentators on *MC,* but one that impinges significantly on the project of historical reconstruction Rushdie undertakes in this book: "Bombay is a city built by foreigners upon reclaimed land," says Rushdie. "I, who had been away so long that I almost qualified for the title of *farangi,* was gripped by the conviction that I, too, had a city and a history to reclaim" (76). Behind the figuring of the "almost . . . *farangi*" (i.e., foreigner) Rushdie, who wishes to re-member and thereby re-claim (his) India, is the figure of Jawaharlal Nehru or, more precisely, the position Nehru ascribes to himself as he undertakes his narrative of the nation, *The Discovery of India:*[9]

> India was in my blood. . . . And yet I approached her almost as an alien critic. . . . To some extent I came to her via the West, and looked at her as a friendly Westerner might have done. (50)

David Lipscomb has convincingly demonstrated Rushdie's critical engagement with a different "dominant history" of India in *MC,* Stanley Wolpert's *A New History of India.* Lipscomb shows how Rushdie has literally incorporated chunks from Wolpert's text in order to unmake "a discourse that claims absolute forms of knowledge, the discourse of Western historiography" (164). The intertextual reference to Nehru's *Discovery* is more oblique but invoked as well to interrogate, if not altogether dislodge, the "truth" of this prior narrative of nation and its representation of India. With this difference, however: As his recent re-visiting of the same terrain in *The Moor's Last Sigh* demonstrates, Rushdie shares and is even emotionally invested in the secular, inclusive vision of the subcontinent's cultures and peoples that Nehru articulates, so his interrogation of *Discovery* is at least to some extent a self-reflexive critique of his own investment in this vision.

Discovery, it is important to note, is viewed as an exemplary instance of Indian bourgeois, nationalist history, and its dominant, virtually canonical status is a function of the power that accrued to Nehru initially as a major nationalist figure involved in anti-colonial insurgency against the British and later as the first prime minister of independent India. Read alongside or, better, through the lens of *MC, Discovery* both overlaps in some of its procedures and differs significantly from Rushdie's representation of India in *MC. MC*'s difference from *Discovery* signals not only a difference that is a product of historical moment and agenda, but also, and more importantly, what turns out to be an overarching critique of the India that Nehru, in part, brought into being.[10]

Like Rushdie's after him, Nehru's narrative is invested in reclaiming a past, but unlike Rushdie's *MC,* Nehru's *Discovery* is undertaken with a view to disclosing India's resilience and (continuing) viability as a nation, a "unique continuity of tradition and cultural life" in which "one hundred and eighty years of British rule" are relegated to the status of "one of the unhappy interludes in her long story" (51). Nehru brings some self-consciousness to his undertaking when he remarks on his perspective and agenda and how these impinge on his narrative. He concedes, for example, that "all of us, I suppose, have varying pictures of our native land and no two persons will think exactly alike" (62). (Compare this with Saleem's "there are as many versions of India as Indians" [323].) Nehru's "picture" is fueled by what he calls his "nationalism." But he asserts simultaneously that this nationalism is "inevitable" for a subject people and thus is not restricted to him alone:

> My reaction to India thus was often an emotional one, conditioned and limited in many ways. It took the form of nationalism. . . . But nationalism was and is inevitable in the India of my day . . . For any subject country national freedom must be the first and dominant urge. (52)

While Nehru's intermittent self-consciousness is somewhat compatible with Saleem's (and Rushdie's) constant awareness of perspective as it impinges on and constructs the (historical) past one reclaims, what is not compatible, and thus becomes the site for interrogation, is the primarily positivist drive of Nehru's narrative, marked by the sense that "the long panorama of India's history" lies before him to record passively and disseminate. What gets overshadowed or even lost in this move is Nehru's initial perception of "condition[s] and limit[s]"—contingencies, that is, that frame his narrative. Thus, when read through the lens of *MC,* in which Saleem's presumption of totalizing, whole sight is constantly undermined, *Discovery*'s (and other traditional, nationalist histories') evasion of the contradictory drives underpinning its narrative authority becomes available for critical evaluation.

Discovery's evasions are often mobilized in the service of a nationalist agenda that must insist on a unified nation state in whose interests anticolonial insurgency can be undertaken and whose unified support, in turn, is necessary for the success of such insurgency. Written, then, with a view to shaping this unity, *Discovery* nevertheless asserts this unity as an already accomplished fact. When *MC* reveals this unity as spurious, focusing instead on an India riven by all sorts of conflicts and contradictions literalized in the disintegrating body and spirit of its central protagonist, it reveals the gap between aspiration and achieved "fact" that Nehru's narrative covers up. In so doing, it calls into question not only the method through which Nehru's truth claims are presented, but the end product of this method as well.

One could say that these contradictions in Nehru's narrative are necessitated by the exigencies of his historical moment and purposes. Driven by the need to free India from the British, and committed simultaneously to elaborating the nation-state that will follow from this freedom, Nehru can hardly afford to bring attention consistently to the contingency of either his account or his vision. But the disclosure of contradictions is merely grist for the mill of a more ambitious critique of Nehru's India that *MC* undertakes. Nor do the limits imposed by historical moment and agenda figure necessarily in this critique. The "disappoint[ment] in the fate of India, the bitter[ness] at broken promises, the humiliat[ion over India's] inability to discard the corruption of the Raj" (Barnett 95) that *MC* records is a disappointment, bitterness, and humiliation that can be laid at least in part at the door of Nehru, who, as one of the architects of post-independence India, is responsible for the nation that came into being. This awareness redounds darkly on the narrative, which sets out to elaborate, if not in all its details, the nation India must become.

3.

Through the dugdugee man Lifafa Das's peepshow, Saleem, Rushdie's protagonist in *MC,*[11] provides an important gloss on his method for representing India. (*MC* is full of such self-reflexive gestures, "wearing" as it does, KumKum Sangari notes, "its technique on the sleeve" [238]). Emphasizing sight ("Dilli Dekho," "Come see Delhi" [83]) and thus perspective and its concomitant, selectiveness, Lifafa Das's peepshow includes:

> pictures of the Taj Mahal, the Meenakshi Temple, and the holy Ganges; but as well as these famous sights . . . more contemporary images—Stafford Cripps leaving Nehru's residence; untouchables being touched; educated persons sleeping in large numbers on railway lines . . . even a newspaper photograph . . . of a fire at the industrial estate. Lifafa Das did not believe in shielding his audiences from the not-always-pleasant features of the age. (84)

Similarly, Saleem's India—the episodes, peoples, cultural forms that mark its pre- and post-independence life—is presented via forms that mimic snapshots, movie stills, pictures, *tableaux vivants.* Organized through this eclectic mix, in which each snapshot/movie still/picture/*tableau vivant* resonates with significance the reader must often draw out, Saleem's narrative eschews linear connections. The hand that puts together these discrete images and the frame that marks the boundaries of each representation are constantly in evidence, underscoring their (interested) constructedness. What is also in evidence is the highly stylized nature of Saleem's narrative, in which movement within each frame is nevertheless contained by the stasis implied by the act of framing as it seeks to capture the singular, resonant image.

In this section I analyze a selection of framed, stylized representations that recycle familiar colonialist and nationalist[12] presentations of India, with a view to disclosing Saleem's subversion of their self-evident naturalness or truth.

Saleem begins his narrative of nation with a familiar set of oppositions that govern these prior representations: "West" versus "East," modernity versus tradition, and so on. These oppositions appear in a series of shots that introduce the recently returned, Western-educated Aadam Aziz, Saleem's grandfather,[13] and Tai the boatman, Aadam's childhood companion, loved by Aadam for his "endless verbiage," "magical talk" (10), and the (now changed) interactions between the two. The description of Tai that stays etched on the reader's mind is the one in which Saleem invokes a recognizable *topos* of colonial and subsequently nationalist representations of the subcontinent when casting Tai as someone who, if not outside history, is certainly beyond its touch. "A quirky, enduring, familiar spirit of the valley" (10), Tai is presented as a veritable incarnation of the changeless, prehistoric traditionalism associated with the "East": "Nobody could remember when Tai had been young. He had been plying the same boat, standing in the same hunched position . . . forever" (9). Saleem repeats this *topos* later in his narrative when he reconstructs the history of Bombay:

> The fishermen were here first. Before Mountbatten's tick tock . . . beyond Dalhousie and Elphinstone, before the East India company built its Fort, before the first William Methwold; at the dawn of time . . . in this primeval world before clock towers, the fishermen—who were called Kolis—sailed Arab dhows. . . . One day in 1633, an East Indian company officer named Methwold saw a vision. This vision—a dream of British Bombay, fortified, defending India's west against all comers—was a notion of such force that it set time in motion. History churned ahead. (105–106)

In the colonialist script, India's prehistoric/primeval identity is assessed negatively as the reason for India's justified subjugation because it renders India (and Indians) historyless, outside time, and thus outside the forces that can produce change. This identity, moreover, is set off against the West's dynamism and access to history, which is presented as something the West in a sense sets in motion. In nationalist accounts, the same *topos* of India as prehistoric/primeval has a quite different, positive resonance. In the nationalist agenda elaborated, for example, by Gandhi, changelessness, which signifies India's (continuing) prehistoric identity, also represents India's *active* refusal to change in response to the disruptions caused by colonial occupation; changelessness thus is recuperated as a mode of resistance against colonial rule. Similarly, Tai's refusal to bathe is recorded as "a gesture of unchangingness in defiance of the invasion of [Aadam's] doctori-attaché from Heidelberg" (25–26). Quite apart from reflecting oppositions deployed by colonial and anti-colonial nationalist discourses, Aadam Aziz and Tai reflect as well an opposition *within* official Indian nationalist thought between Nehru's and Gandhi's disparate visions about the kind of struggle that should be waged against the British and the kind of nation India should aspire to become after independence. Nehru "was anxious to change India's outlook and appearance and give her the garb of modernity" (*Discovery* 50). His vision of a secular India, to which end the narrative of *Discovery* is deployed, was a vision of a modern, Westernized nation. It was only through acquiring and thereafter claiming a full-blown modernity and its correlate, an "enlightened," "scientific" rationality that needs must jettison a substantial portion of the country's "past heritage and traditions" that India's destiny, according to Nehru, would be realized. Gandhi's *Hind Swaraj*, "one of the first, and perhaps fullest [attempts]," says Partha Chatterjee, of "a systematic exposition of his ideas on state, society and nation" (*Nationalist Thought* 85), is unequivocal in its condemnation of "modern civilization" (*Hind Swaraj* 38). Far from endorsing India's acquisition of modernity, Gandhi recommends a return to what he views as the past of a self-sufficient, traditional India, where "we managed the same kind of plough as existed thousands of years ago. We have retained the same kind of cottages . . . and our indigenous education remains the same as before" (55). In the Gandhian script, India's unchangingness is assessed not as a handicap, but as a strength:

> It is a charge against India that her people are so uncivilized, ignorant, and stolid, that it is not possible to induce them to adopt any changes. It is a charge really against our merit. What we have tested and found true on the anvil of experience, we dare not change. (54–55)

Thus, far from "impertinently excis[ing]" Gandhi from his narrative of nation, as Brennan complains, in his *Salman Rushdie and the Third World* (84), Rushdie's Saleem makes Gandhi's vision of an independent India available to us through Tai.

In these "shots," Saleem's re-staging of both colonialist and nationalist self-descriptions and representations of the subcontinent that themselves deploy the same *topos* but assign entirely different meanings to it in effect underscores how any single representation is open to radically differing assessments when it is deployed in response to radically different ideological agendas.[14] Re-play here acquires its critical force through covert juxtaposition of these two contradictory assessments of the same *topos,* assessments that in effect disrupt the authority of any single deployment of that *topos* as the "truth" about the subcontinent.

One could, of course, view this juxtaposition of contradictory assessments with regard to the same *topos* as an attempt to underscore the indeterminacy that haunts any attempt at interpretation. But a focus on Rushdie's particular ideological investments would most likely lead us in a different direction.

Consider, for example, Rushdie's bitter (albeit humorous) diatribe in "Outside the Whale" against contemporary British re-plays of the "novel of empire" (Brennan's phrase)—on the screen (in David Lean's rendition of Forster's *A Passage to India*) and in fiction that has been re-made into TV serials (Paul Scott's *The Raj Quartet* and M. M. Kaye's *The Far Pavillions*). This diatribe raises some very interesting questions about a strategy that Rushdie himself uses and about Rushdie's own re-play in *MC* of certain details in Forster and Scott.[15] Let us remember that "Outside the Whale" (1983) was published three years after *MC.* Rushdie's critique of contemporary British re-plays, then, may be seen as an attempt to clear space for his own representations of the subcontinent, especially since he re-plays at least one of the works he criticizes, Scott's, and notes how both Scott and Kaye draw upon Forster's *A Passage to India* and Kipling's *Kim,* which reappear in *MC* as well. However, alongside Scott's and Kaye's works and the TV serials adapted from them, and with reviews of *MC,* some of which compare his work with Scott's, already in circulation, Rushdie's critique can equally be seen as the means by which he wants his readers to distinguish his own project from that of Scott and Kaye which he unequivocally characterizes as "fake portraits inflicted by the West on the East" (126).

In order to distinguish his own work, implicitly, if not explicitly, Rushdie raises the issue of literary merit as an index of individual talent: Both *The Far Pavillions* and *The Raj Quartet*

> are founded on identical strategies of what, to be polite, one must call borrowing. In both cases, the central plot-motifs are lifted from earlier and much finer novels. . . . But because Kaye and Scott are vastly inferior to the writers they follow, they turn what they touch to pure lead. (*Outside the Whale* 127)

He follows this indictment with what I consider a more formidable, because more fundamental, distinction between his re-plays and those of

Scott and Kaye, which in turn implicitly forbids us to see their "borrowings" as re-plays in the way, for example, I have defined Rushdie's use of this strategy. He situates Scott's and Kaye's projects within the context of—indeed, sees them as embodying—"the recrudescence of imperialist ideology" in present-day Britain (129). Far from displacing, unsettling, or interrogating prior hegemonic representations, Scott and Kaye uncritically repeat in order to restore the ideology that fueled these representations in the first place.

To grasp fully the distinction suggested by Rushdie's statements, it is useful to turn to Gayatri Chakravorty Spivak's comment that the "imperialist project" was

> the worlding of a world on supposedly uninscribed territory . . . a worlding that was also a texting, a textualizing, a making into art, a making into an object to be understood. (*Post-Colonial Critic* 1)

Rushdie's, unlike the imperialist's, is a post-colonial or post-imperial project, one that, according to Gyan Prakash, "exists as an aftermath, an after—after being worked over by colonialism" ("Postcolonial Criticism and Indian Historiography" 8). For Rushdie, who sees his work as involved in the effort to "draw new and better maps of reality" ("Outside the Whale" 137), the fact of his coming "after" is an invitation to examine critically and interrogate that which has gone before. Not only does he not want to repeat uncritically the actions, discursive or otherwise, of the colonizers, who proceeded as if the earth they territorialized was in fact previously "uninscribed," but Rushdie's re-play of the dominant also suggests that his (or anyone else's) critique cannot simply leap over or clearly situate itself outside what is being critiqued, which in his work includes prior structures of domination—structures that secured domination as much as reflected it. Instead, critique entails inhabiting those structures once again so that the contradictions, silences, and exclusions through which they established their authority and made themselves invincible are revealed. Intimately inhabiting the source of one's critique is also a way of signaling, as my reference to Spivak's comments on "critique à la deconstruction" makes clear (see note 7 in this chapter), one's own and therefore one's work's imbrication in that source.[16]

The example from *MC* that I now turn to demonstrates why re-play in Rushdie is never uncritical repetition, never an attempt merely to restore a prior hegemonic representation. Although not directly borrowed from a "novel of Empire," yet depicting scenes that could easily have come from any one of such works and that certainly evoke a cluster of ideas occupying a revered place in British self-descriptions (and mythology) about their orderly and civilized departure from the subcontinent, the Methwold episode in *MC* describes Methwold's elaborate preparations for the transfer of real estate to carefully selected Indians and his subsequent departure for England. Both the preparations for transfer and departure are linked, at the insistence of Methwold himself, with those of his country's:

> My notion . . . is to stage my own transfer of assets, hand everything over absolutely intact: in tip-top working order. (111)

Later, on the day of his departure:

> [A]t four o'clock that afternoon William Methwold drives up to the two-story hillock in his black 1946 Rover. He parks in the circus-ring between his four noble villas . . . he takes up his position in the exact centre of the circus-ring. Rose in lapel, cream hat held stiffly against his chest, centre-parting glinting in the afternoon light, William Methwold stares beyond Breach Candy's map-shaped pool, across the golden four o'clock waves, and salutes; while out there, above the horizon, the sun begins its long dive towards the sea. (130–31)

In this *tableau vivant,* in which the Englishman is presented (once again) at the center of affairs, orchestrating the details of his departure as he hands over his property to a group that symbolizes the neo-colonial elite, the hurly-burly of history is stilled, the hard-won and bitterly fought struggles for independence silenced, the blood-letting that preceded and accompanied the partition of the subcontinent into two nations dismissed earlier and outside the frame of this scene by Saleem's willful "I shall not describe . . . I shall avert my eyes . . ." (130).

Although Methwold, in Saleem's rendition, reveals considerable self-consciousness with respect to the way in which he stages his departure, which parallels that of his country's, the same cannot be said of colonialist renditions or, according to Rushdie in his "The New Empire Within Britain," of present-day "white Britons'" assessment of the departure of the Raj.[17] They do not see British representations of the events leading up to the departure and the departure itself as deliberate, motivated staging, at odds with what actually happened and fueled, for example, by the desire to present the formal independence of India from the British as a gift given, an instance of *noblesse oblige* ("hand over everything absolutely intact: in tip-top working order"), not as something the Indians actively fought for and won.

Nevertheless, Methwold's orchestration of his departure, as Saleem presents it, in its intense focus on himself completely disregards the violent events that marked the lives of Indians and that were happening as the British were preparing to leave. This exclusion can be seen as motivated by the desire not to bring attention to events that were at least in part the product of British colonial policies. Like Methwold, Saleem, narcissistically focused on himself, dismisses these events even as he explicitly draws attention to the motivated nature of his excision of them: "Narrow-minded? Well, perhaps; but excusably so, in my opinion. After all, one is not born everyday" (130). In doing so, Saleem draws attention to the mesh of self-interest and desire that underlies all (historical) narratives but is most often normalized, clandestinely rendered natural or self-evidently true.

Of course, Rushdie's two novels about the Indian subcontinent suggest that such self-interest and will to power are not restricted to dominant colonialist representations alone. They characterize nationalist descriptions as well. What's more, his own representations are implicated in something comparable. "Draw[ing] new and better maps of reality," after all, is not only a project that re-envisions reality; it also entails *imposing* what one sees as a better vision of reality on others. However, the one thing consistently absent in colonialist and nationalist representations but persistently present, almost to an excruciating degree, in the accounts by Rushdie's narrators is a self-reflexiveness about the perspectives from which their narratives are being told. Thus, while colonialist and nationalist accounts proceed as if they were unproblematically and self-evidently reflections of the realities of the subcontinent, Rushdie's narrators anxiously examine and display their motivations, elaborate defenses for their judgments, worry about their fallibilities and inadequacies in order to emphasize their accounts' mediatedness:

> Am I so foregone in my desperate need for meaning, that I'm prepared to distort everything—to re-write the whole history of my times purely to place myself in a central role? (*MC* 198)

> I am literally dis-integrating. . . . This is why I have resolved to confide in paper, before I forget. (We are a nation of forgetters.) (*MC* 37)

> *Outsider! Trespasser! You have no right to this subject! . . . Poacher! Pirate! We reject your authority. We know you, with your foreign language wrapped around you like a flag: speaking about us in your forked tongue, what can you tell us but lies? I reply with more questions: Is history to be considered the property of participants solely?* (*Shame* 23)

> I have learned Pakistan in slices. . . . However I choose to write about [it], I am forced to reflect the world in fragments of broken mirrors. . . . I must reconcile myself to the inevitability of missing bits. (*Shame* 70–71)

By foregrounding his narrators' anxieties and vulnerabilities, Rushdie is asking his readers to ask questions about the stories they tell, their emphases, modes of description, inclusions, and exclusions, which are shown as filtered through and motivated by his narrators' quotidian and other needs, their cultural and ideological investments. To the extent, moreover, that Rushdie's narrators repeat not only the method but sometimes also the content and evaluation of colonialist and nationalist descriptions, the questions we must ask of his narrators extend to those we must ask of these prior representations as well.

4.

In my discussion of the ways in which Saleem describes Methwold's departure, I have argued that the strategy of re-play is mobilized in the interests of re-staging an actual historical event—the departure of the Raj—to unsettle and open up for inspection British self-descriptions of a departure that, despite being pressed into the service of mythologizing, was presented as, and acquired the force of, the "real." To the extent that Rushdie's narrator underscores such self-descriptions as mythic, Gyan Prakash is right when he sees *MC* as an instance of what he calls "Post-Orientalist Histories" because of what it discloses about the "fable-like character of real history" ("Writing Post-Orientalist Histories" 405). Given some of contemporary historiography's interest in how, why, and what "facts" are "isolated . . . and served up as 'history'" (Pandey xi), which emphasizes not only the question of perspective but also the integral role interpretation plays in the reconstruction of past events, contemporary historiography's overlap with the literary, with what Hayden White calls "verbal artifacts" (82), has become a productive conjuncture that enables one to interrogate traditional historiography's claims to objectivity and settled truth. Of course, Rushdie works in full awareness of this conjuncture not only by casting his literary efforts in *MC* and *Shame* as historical accounts that include significant events and personages from the pre- and post-independence histories of the subcontinent, but also by inserting comments on definitions of history and on the task of historical reconstruction.

Moreover, in keeping with some of recent historiography's interest in what traditional historiography and official accounts of the nation leave out, Rushdie's narrators elaborate upon various exclusions, noting as well the ideological imperatives that govern acts of exclusion. Most often, Rushdie's narrators bring to light exclusions by going over again a set of historical events, re-viewing what "[h]istory books, newspapers, radio programmes tell us" (*MC* 497) in order to disclose, through personal memories and experiences (the narrators' own and others'), a different or alternative truth or set of truths, revealing thereby the deliberate silences or suppressions in dominant accounts. Saleem does this when he describes the civil war between the east and west wings of Pakistan, challenging both official Pakistani and Indian accounts of the civil war and the creation of Bangladesh (415–54). He does it again when he recalls the 1975–1977 emergency rule in India, when the then-prime minister of India, Indira Gandhi, through a series of draconian measures, violated the civil rights of Indians (461–531). He recalls official accounts alongside his own alternate ones not only because he wants to reveal the discrepancies between the two, but also because he wishes to demonstrate how discovering and recovering exclusions is itself a function of going over again—a function, that is, of re-play.

Shame also is invested in unearthing exclusions and exposing them. In this work, moreover, exclusions provide both the content for various stories the narrator tells and the method, the conceptual and interpretive framework through which the "facts" of Pakistan's official culture and sociopolitical arrangements are selected and "served up as history." In *Shame,* the narrator's historical reconstruction is in fact presented as mythic, a "fairy-story" whose power to disrupt official accounts of Pakistan exists in direct proportion to our recognition, in turn, of the latter's mythic character as well.

Women and their perspectives constitute what the narrator of *Shame* considers one of the central exclusions of Pakistan's male-dominated history, culture, and sociopolitical arrangements. Inviting a comparison between what he construes as women's marginalized position and his own peripheral status,[18] the narrator, an immigrant in England and an intermittent visitor to Pakistan, "refracts" the official narrative of Pakistan "through the prisms of its reverse and 'female' side" (189) to forward evaluations whose emphases (and investments) are quite different from those implied in official narratives.

It is important to emphasize here a distinction implied in the quote above if we are not to misunderstand Rushdie's project, its assumed limits that nevertheless limit how women get written about and into the alternative narrative of nation *Shame* offers. Although the exclusion of women and their perspectives from official accounts of nation enable *Shame*'s alternative account, it is through the narrator and his take on what Pakistan's history, its cultural and sociopolitical arrangements, would look like if they were "refracted" through the position he specifies as "female" (and as the "reverse" of "male") that we get this alternative account. *Shame*'s narrative, in other words, comes to us through a narrator who appropriates what he construes as a "female" positionality, a construction that proceeds from his understanding of how a male-dominated society construes the female as weak, ineffectual, and powerless while aggrandizing all power, strength, and agency for (and in the interests of) the male.[19] This understanding leads to (or better, supplies the grounds for) the narrator's bitter denunciation of what he views as Pakistan's brutally gendered cultural and sociopolitical arrangements. Given these, the narrator finds it hardly surprising that Arjumand Harappa regrets her gender. Indeed, in a "Man's world," her access to public space and power networks dictates that, following her father's advice, she "rise above her gender" (136), that she, quite literally transform herself into a man, which is what she proceeds to do:

> When her breasts begin to swell, she will bind them tightly in linen bandages, so fiercely that she blushes with pain. She will come to enjoy the war against her body, the slow, provisional victory over the soft, despised flesh. (136)

This refraction "through the prisms of its reverse and 'female' side" recuperates (and re-plays) the official male-dominated history of Pakistan not as a "heroic" narrative, but as a farcical "saga of sexual rivalry, ambition, power, patronage, betrayal, death, and revenge" (189). This account locates all violence, shamelessness ("Wherever I turn, there is something of which to be ashamed" [22]), and corruption in the "male'" and as proceeding from "male" prescriptions about "honour" and "shame" or "dishonour." It sees the political and cultural map of Pakistan marked by "brothers [who] have done the worst of things to brothers" (308); by the betrayal of "sons" against their "fathers," of protégés against their mentors (Haroun kills his father; Raza Hyder turns against Iskander Harappa, who, by his own reckoning, "*made*" Raza, "*from darkness into light, from nothingness into somethingness*" [254; emphasis in original]). And it reveals women burdened by their oppression and by the violence directed against them (including violence that they have internalized, as, for example, Arjumand's transformation of her "femaleness" into "maleness" demonstrates). Women "escape" this oppression and violence by going mad (Bilquis), by committing suicide (Naveed "Good News" Hyder), or by externalizing the shame and violence passively absorbed (Sufiya Zinobia, *Shame*'s nominal heroine, who, as first-born daughter who should have been a son, is "a miracle-gone-wrong, a family's shame made flesh" [150]).

The narrator's appropriation of a "female" positionality whose definition proceeds from dominant constructions of "male" and "female," albeit working with an awareness of these constructions' brutalizing effects, especially on women, has a curious effect: it ends up confirming, instead of simply rendering suspect, dominant constructions of women as weak, ineffectual, and powerless. This, it can be argued, is the inevitable risk (perhaps even an unavoidable blind spot) in appropriations of the dominant. In other words, what is appropriated (even for counterhegemonic purposes) inevitably implicates the person (and the work) undertaking the appropriation. More significantly, I think Rushdie's narrator's overdetermined vision of Pakistan as a nation driven, for the most part, by a deep and fundamental violence, is rendered specific through, and exemplified by, the violence he views as endemic to the position women occupy within Pakistan's dominant (male) culture. The suppressions and hypocrisies that the narrator associates with this "male" culture, in other words, accrue from and dialectically reproduce this violence. As such, his account of women's (impossible) position is crucial to a critique—that is also the basis for his alternative account of the nation—invested in elaborating the chicaneries and shamelessness of the ruling (male) elite.[20]

The female character who seems to escape the fate of other women in *Shame* is also one who offers an alternative narrative that closely replicates (or perhaps enables) the narrator's own focus on "unspeakable things" (210) perpetrated by the ruling elite. It's tempting, therefore, to see the positionality specified as "woman's" or "female" in *Shame* as receiving its fullest treatment in Rani Harappa's "eighteen shawls of memory" (210), an "embroidered" account of her husband, Iskander Harappa (a thinly disguised stand-in for Zulfikar Ali Bhutto, the former prime minister of Pakistan and the father of Benazir Bhutto, whose own tenure as prime minister preceded that of Zia ul Haq, who in *Shame* appears as Raza Hyder). An implacably oppositional account, Rani Harappa's shawls, entitled "The Shamelessness of Iskander the Great," document "unspeakable things" (210). The narrator, following the Rani's "embroidery," represents them as a series of allegorical "pictures" that distill the horrors perpetrated during Iskander's rule:

> The shawls of international shame, Isky grovelling at primrose Chinese feet, conspiring with Pahlevi, embracing Dada Amin; the eschatological Iskander riding the atomic bomb. . . . Iskander and the Death of Democracy, his hands around her throat, squeezing Democracy's gullet. . . . the shawl of hell [depicting] . . . the separatist movement . . . there it all was in scarlet, scarlet, nothing but scarlet, what he did for the sake of no-more-secessions, in the name of never-another-East-wing . . . men without genitals, the sundered legs. (212–13)

An indictment on a grand scale, Rani Harappa's narrative (like the narrator's) describes those elements of actual events and characters that official or dominant accounts often sweep under the carpet. But Rani Harappa's narrative, though forceful and persuasive within the framework of the narrator's own interpretive take on Pakistan's ruling elite, is also contradicted by her daughter Arjumand Harappa's perception of her father as a veritable saint. Moreover, it is only one—albeit a valuable—way to look at Iskander's reign and is hence marked by the limits that characterize any single account: "Every story one chooses to tell," remarks Rushdie's narrator at one point, "is a kind of censorship; it prevents the telling of other tales" (72–73). More important, its ability to disrupt prevailing conceptions of Iskander's greatness is undermined insofar as it documents "unspeakable things nobody wanted to hear." Even the Rani seems to recognize this when she chooses her "newly powerful daughter, Arjumand"—the person least likely to be persuaded of the force or legitimacy of Rani's views of Iskander and therefore least likely to give them wide currency—as the recipient of her gift of shawls, which tell an alternative tale of Iskander's rule.

Even though the narrator disseminates the Rani's counternarrative within the pages of his own with a view, like the Rani's, to providing a different account of the nation Iskander made, through his (and the Rani's) recognition of the precarious—almost negligible—access this narrative has to the public space defined and constricted by dominant ideology, Rushdie seems to be making a more general point about the efficacy and power of alternative narratives (including his own) to transform existing social, political, and cultural arrangements. Dominant accounts are powerful precisely because their force or legitimacy depends upon the ease with which they can be disseminated, an ease that undoubtedly depends upon their connections with those in power. But this ease also accrues from the fact that what they offer in terms of normative judgments about events, people, culture, politics, and so on has been internalized by large segments of people, including those left out of these accounts' preferred self-definitions, towards whom they are directed. Thus alternative or oppositional accounts (and the assumptions that undergird them), like the Rani's and in some respects the narrator's, often at odds with what most people would recognize or endorse, must perforce lead a shadowy life, overcome obstacles that include modes of containment through which dominant accounts incorporate or, more overtly, censor them.

To recognize this, however, does not entirely denude the Rani's or the narrator's alternative renditions of official history of their efficacy or of their ability to make a difference. The suppressed or excluded truths they rehearse are important to rehearse precisely because they challenge the authoritative, seemingly invincible truths of dominant accounts. Furthermore, insofar as dominant accounts (like the dominant ideologies that underpin them) cannot commandeer all social, cultural, and political space for their preferred self-definitions, alternative accounts can (and do) lodge themselves in the "vacant" spaces, limited though these might be, at the very least to interrupt, qualify, provoke uncertainty about the force or legitimacy of normative ways of seeing and inhabiting, in this instance, one's nation and culture.

Both *MC* and *Shame* proceed from the awareness that dominant representations, values, and ideas cannot be simply shrugged off as irrelevant to the task of mapping alternative ways of seeing, alternative structures of knowledge, not the least because the dominant's power accrues from its ability to persuade people of its legitimacy. The task that critique undertakes, then, is that of disclosure; it sets out to make visible the methods through which the dominant acquires legitimacy and persuasive force. As a gloss on his work, Rushdie's assertion about the "genuine need for . . . books that draw new and better maps of reality, that make new languages with which we can understand the world" acquires a quite specific meaning: Envisaging not so much the need to deliver a "new" world and language that will replace and/or transcend the contaminations of the one we already inhabit, Rushdie's narratives assert more the need to provoke a process of "unlearning"—an "unlearning" that is, in Raymond Williams's fine phrase, "of an inherent dominant mode"—that begins with a clear understanding of the processes by which we have wittingly or unwittingly learned the dominant.

5.

In a series of interviews and essays given or written around and between the publication of *MC* and *Shame,* Rushdie unequivocally privileges the "migrant sensibility"—particularly that of the imaginative writer/creator—as the source of, even the site for, an adversarial perspective on what is construed as normative (oppressive and oppressing) ways of inhabiting one's world. In his essay on Terry Gilliam's film *Brazil* ("The Location of Brazil"),[21] Rushdie, "borrowing," perhaps not surprisingly, from Romantic self-mythologizing about the role of the poet/creative artist, asserts the "triumph of the [poet/artist's] imagination," the "power of dream worlds" to oppose "dark reality," including the power of politicians and government functionaries whose business it is to suppress uncomfortable truths. Ascribing to the creative imagination the power to construct and make available a better world, Rushdie also ascribes to it the ability to break through "conventional habit-dulled certainties" that make people think "resistance is useless." Thus, imagination can transform and thereby instill us with "confidence in our ability to improve the world"; imagination is the "only weapon with which reality can be smashed, so it may be subsequently reconstructed" (52–53). And this imagination, this ability to re-make the world, render it anew, is most especially a possession of the migrant's.

> Migrants must, of necessity, make a new relationship with the world because of the loss of familiar habitats. . . . [T]he location of *Brazil* is cinema itself, because in cinema the dream is the norm . . . this cinematic land of make-believe of which all of us who have, for whatever reason, lost a country and ended up elsewhere are true citizens. Like Terry Gilliam, I am a Brazilian. (53).

Such a privileging of the migrant and of "migrant sensibility" as the authoritative source of an oppositional or dissident vision in Rushdie's jounalistic prose, while of a piece with a great deal of the theorizing about postcoloniality in the West, particularly in the Anglo-American academy,[22] is nevertheless at odds with with the more oblique, deconstructive procedures of Rushdie's fiction, in which singular, authoritative assertions and positions are more or less consistently destabilized and interrogated, and in which, moreover, they are subjected to qualifications and disclaimers that by attending to the ideological mediations of all accounts, assertions, and positions render them more modest, framed by specific rhetorical and political urgencies. Alternative, counterhegemonic accounts and/or perspectives in Rushdie's fiction, even as they seek to undermine the invincibility and power of the dominant are, nonetheless, implicated in (not conceived as outside or the opposite of) the dominant.

It behooves us, I believe, to read Rushdie's assertions about the migrant's (inevitably) oppositional vision along with and through the lens of the deconstructive procedures of his fiction and vice versa. Doing so might enable that form of critical "interruption" that Spivak describes as necessary for an understanding of all forms of representation and knowledge claims,[23] and that is so central to the task of critique in Rushdie's fiction. Doing so might also urge the recognition that Rushdie's unequivocal privileging of the migrant's position vis-à-vis dominant representations and ideologies is a function of his own location and its specific urgencies. This privileging, in addition, is the means by which he empowers himself as not quite an insider, but not quite an outsider either, to write about, and speak on behalf of, the Indian subcontinent.

Part II

Return(s) to the Native Land: The Politics of Cultural Nationalism

For if imperial travelers and colonial writers alike assumed that England was the oikos*—the place of return . . . anti-imperial nationalism came to be propelled by a narrative that insisted on retour as a necessary condition for alternative identities.*

—Simon Gikandi

I speak from a remove in time and space—I have tried to hold your shape and history within me. I keep track of you through advertisements and photographs. Through The Harder They Come *and "Natty Dredd." By watching a group of apple pickers stroll through a Massachusetts hilltown. Through shots of graffitti on the walls of Tivoli Gardens. My homeland. My people. . . . I wonder if I will ever return. . . . I am afraid my place is at your side. I am afraid my place is in the hills. . . . I bear in my mind that you with all your cruelties are the source of me, and like even the most angry mother draw me back.*

—Michelle Cliff

Well, nationalism is such a powerful term for us (Africans) because of what we have been through as a people and are still going through. Over the last five hundred years, African people have been under all kinds of onslaught—physical, mental, emotional. It seems to me that whatever is left for us to recoup cannot be done unless we see ourselves as a people, as a nation. When I say African nationalism, I am also using the term to embrace the global African world. . . . Maybe then the better term for me to use is PanAfricanism.

—Ama Ata Aidoo

Chapter 3

Reversing Europe's Gaze: Ama Ata Aidoo's Oppositional Art

How do colonial subjects write their identities within the cultural totality established by imperialism? If power still resides in the metropolitan center, even after empire, what is the meaning of a reversed gaze at the culture of [neo-imperialism]? what vehicles do the colonized use to represent [it]?

—Simon Gikandi

In an underdeveloped country an authentic national middle class ought to consider as its duty to betray the calling fate has marked for it, and to put itself to school with the people: in other words, to put at the people's disposal the intellectual and technical capital that it has snatched when going through colonial universities. But unhappily we shall see that very often the national middle class does not follow this heroic, positive, fruitful, and just path.

—Frantz Fanon

I.

In 1952, Frantz Fanon, viewing the prospect of "millions of [colonized black] men who have been skilfully injected with fear, inferiority complexes, trepidation, servility, despair, abasement,"[1] attempts, through his analysis in *Black Skin, White Masks,* to eliminate "once and for all" the "poison" of internalized racism and black self-hatred. "I propose," Fanon avers, "nothing short of the liberation of the black man from himself" (8). His "theme being the disalienation of the black man," Fanon hopes his "study" will "enable the man of color to understand, through specific examples, the psychological elements that can alienate his fellow Negroes" (79). Almost two decades later, addressing a different, albeit not radically dissimilar, historical conjuncture and geographical site, Ama Ata Aidoo, confronted with a comparable prospect of the colonized, deracinated African, also attempts through her work—particularly *Our Sister Killjoy; or Reflections from a Black-eyed Squint* and her collection of short stories, *No Sweetness Here,* which I will examine in this chapter—to restore the health and well-being of her beloved Africa and fellow Africans.[2]

Fanon haunts Aidoo's work, both through specific echoes from *BSWM* (which I will have occasion to refer to later) and more generally through his prescient critique of the national bourgeoisie, which, instead of committing the class-suicide Fanon and other revolutionary leaders recommended,[3] consolidated its position by helping to prolong imperial domination through neo-colonialism. Sharing their object of critique and their belief in the pedagogical power of analysis and exposure,[4] Fanon and Aidoo share as well a Manichean conception of their respective worlds with the colonizer and the colonized locked in implacable opposition to each other.[5] However, despite their belief in decolonization as a form (and product) of national liberation, they differ with regard to the emancipatory potential each grants to nationalism as an ideology and form of consciousness.

For Aidoo, nationalism embodies by far the most empowering stance against and resistance to imperialism and its ongoing presence through various forms of neo-colonialism. Responding to Molara Ogundipe-Leslie's comment that "the female African writer should be committed in three ways: as a writer, as a woman, and as a Third World person," Aidoo asserts the centrality of Africa to her concerns, and her "need . . . to be an African nationalist"[6]:

> I don't deny that we belong to a larger non-northern world and the dynamics that operate in a situation like that, but find my commitment as an African, the need for me to be an African nationalist, to be a little more pressing. It seems there are things relating to our world, as African people, which are of a more throbbing nature in an immediate sense. (Qtd. in Adeola James 14)

Conversely, in "Pitfalls of National Consciousness," Fanon, viewing the workings of nationalist ideology and consciousness primarily as products of the national bourgeoisie's self-interest, is less convinced of nationalism's emancipatory potential; rather, he offers a scathing critique of some of its significant operations. For example, he finds that the national bourgeoisie's attempts to "wave aloft the notion of nationalisation and Africanisation of the ruling classes" is "more and more tinged with racism," a "racism of defense," he clarifies, "based on fear" (126, 132). Neil Lazarus distinguishes between "two kinds of nationalist ideology" in Fanon "in the context of anti-colonialism": a "bourgeois nationalism" versus "a liberationist, anti-imperialist, nationalist internationalism." "The former," according to Lazarus, "represented the interests of the elite indigenous classes and was aimed at the (re-)attainment of nationhood through means of capture and subsequent appropriation of the colonial state"; whereas "the latter . . . was represented in the Algerian arena by the Front de Libération Nationale . . . to whose cause Fanon dedicated himself." Following Anouar Abdel-Malek, Lazarus characterizes the latter as a "'nationalitarian phenomenon,'" which "'has as its object . . . uprooting in depth the positions of the ex-colonial power—the reconquest of the power of decision in all domains of national life'" ("National Consciousness" 198). Whether or not one can distinguish the latter phenomenon from the former with ease and as sharply as Lazarus does is open to debate. But one significant dimension that most nationalisms share, Aidoo's included, but that Fanon rejects as "vulgar tribalism" (132) is the belief in a singular, essentialist, national identity and identification ("African," in Aidoo's case), with their more or less organic connection with the territory designated as one's nation/home.

Although this difference between Fanon and Aidoo cannot be reduced merely to the difference in their respective locations (spatial, epistemological, historical, and so on), location nevertheless has an important bearing on their respective stances on nationalism. "Authentic" national belonging, such as that assumed by most nationalisms, is not available to Fanon or at least not given the terms of identi(ty)fications he chose as a colonized Antillean, educated in France, practicing psychiatry in North Africa, whose claim to prominence in the annals of revolutionary praxis comes from his intimate connection with and seminal role in the Algerian resistance struggle through the Front de Libération Nationale. Whereas, despite her frequent visits to, and time spent in, the West, away from "home," Aidoo does assume a continuity and an organic connection between Africa and her identi(ty)fications. Born in 1942 in Ghana's central region, Aidoo "acknowledges," as Ketu H. Katrak points out, "that her mother shared a rich heritage of 'talking' stories with her daughter" (136). With a father who, in agreement with Kwame Nkrumah, supported the education of women, Aidoo "in spite of the prevailing bias against schooling girls" (qtd. in Katrak 136) received her education in Wesley Girls' High School at Cape Coast. In her conversation with Adeola James, Aidoo notes how her "parentage" (which includes a paternal grandfather who was "tortured to death in a colonial prison for being 'an insolent African'" [qtd. in Katrak 136]) deeply informed her commitment to the "destiny of our people" (13–14). Her advocacy on behalf of Africa and African subjects is then viewed (by herself and by her readers) as linked intrinsically with her personal family background. So too is her identi(ty)fication as a feminist and a nationalist: "You cannot be a feminist if you are an African," she asserted in the course of her interview with me, "without also being an African nationalist" (124). Here the identity that ties together her roles as feminist and nationalist is, as Aidoo puts it, "African."

Just as Aidoo's privileging of a (African) nationalist identi(ty)fication distinguishes her from Fanon, so too her view that this identi(ty)fication constitutes the most authentic and pure form of resistance to colonialism and neo-colonialism distinguishes her from the two writers, C. L. R. James and Salman Rushdie, whose modes of resistance I have looked at so far. James and Rushdie, I have argued, articulate resistances that emerge from within the folds of dominant colonial ideologies and representations. Their critique, to borrow from Sarat Maharaj's characterization of Black Art, proceeds from an "initial complicity with the material it denounces . . . carv[ing] out its sense of critical distance by first becoming one with it" (15). This critique, I further suggest, is a product of the liminal position—situated at once inside and outside—James and Rushdie perceive themselves as inhabiting with respect to dominant (colonial) ideologies, and which they privilege via the figure of the migrant in Rushdie and diasporic peoples in James. Aidoo's view of resistance as nationalist identi(ty)fication that is autonomous from and uncontaminated by the colonialism and neo-colonialism it resists is a position *apparently* antithetical to the one I have elaborated in the previous two chapters.

Benita Parry forcefully argues for the value of a position on resistance such as Aidoo embodies. Critical of positions that disparage "nationalist discourses of resistance" while privileging "analytical strategies which in focusing on the deconstruction of the colonialist text, either erase the voice of the native [i.e., the colonized] or limit native resistance to devices circumventing colonial authority," Parry is more enamored of a "combative subject position," one that "stand[s] in unmitigated antagonism" to what it resists ("Problems" 35, 34, 30). Parry desires a text of (or position on) resistance that is aggressively "counter-hegemonic," a "possessor and practitioner of an alternative tradition challenging imperialism's authorized system of knowledge" (30). Whether or not we agree with Parry's characterization of the effects of "analytical strategies" she characterizes as deconstructive or those she finds genuinely "counter-hegemonic," insofar as Aidoo's work gives us access to a mode of resistance that Parry would describe as genuinely "counter-hegemonic," and, moreover, one that is quite distinctive in tone and strateg(ies) from the deconstructive practice of James and Rushdie, examining it allows us to grasp the varied, even oppostional strands in positions on (anti-colonial) resistance.

In this chapter, therefore, I analyze Aidoo's work by focusing especially on what Jonathan Dollimore, in a different context, has labeled "reverse discourse"—"inversions" in Aidoo's work of colonial ideologies, representations, and narrative forms. Inversions are an instrumental part of anticolonial nationalism. [7] An inversion, insofar as it reverses or turns about or changes to its exact opposite a belief, value, or form of representing an entity, cannot by definition occupy a space outside that which it inverts. Inversions of colonial ideology and representations in Aidoo's work then could be viewed as existing in contradiction with her desire to have anticolonial resistance occupy a space outside the logic of colonialism. But as I suggest in the introduction, by proceeding as if her critique emerges from a position external to, and not dependent on, what she critiques, Aidoo produces a discourse whose rhetoric and mode of presentation in some instances are quite distinct from those critiques that acknowledge their coimplication in discourses of domination.

It is important to note, however, that although colonial ideologies and representations are a substantial focus of Aidoo's critique, they are grist for the mill of her efforts to persuade the deracinated, Westernized, bourgeois African subject out of his [*sic*] enthrallment with the West[8] so that the task of building a new, genuinely independent Africa can begin.

2.

Ama Ata Aidoo's *Our Sister Killjoy; or Reflections from a Dark-eyed Squint* (hereafter *OSK*) is, among other things, a narrative about a young Ghanaian woman, who, under the auspices of an international volunteer organization, INVOLU, travels to a small Bavarian town in Germany to work in a pine nursery. Reversing the direction of a classic colonial genre, the travel narrative, through which Europe typically represented its soon-to-be or already colonized other, Aidoo reverses as well the gaze that constitutes Europe and its presumed obverse.[9] This time around it's her protagonist's "Black-eyed Squint," not a white, usually male gaze, that functions as the all-seeing eye.

In a journey entitled "Into a Bad Dream," Aidoo's narrator(s) and protagonist[10] represent Europe as a negative, morally inadequate, even reprehensible entity in an opposition that has Africa implicitly representing its obverse—a positive, innately superior, and moral entity. This transposition in values normatively associated with Europe and Africa in colonial narratives (but not only in those) has elicited the charges of a "black exclusiveness" verging on racism, "a revived ethnic negritude," against Aidoo (Brenda Cooper 34, 39). Indeed, insofar as Aidoo locates all positive value in blackness and in an African identity, there is a strong racialized component to her nationalism.[11] But her valorization of blackness/Africa, and her concomitant disparagement of whiteness/Europe is also (and more significantly) an attempt to reverse colonial binaries that presume the superiority of the latter and the inferiority of the former. Reversals such as the ones Aidoo employs are often judged incapable of effecting genuine transformations, trapped as they remain within the logic of the binary terms they invert but do not dismantle. Nevertheless, they are also often viewed as powerful, even necessary weapons in the arsenal of the oppressed. Jonathan Dollimore, noting how "binaries remain fundamental to, and violently active within, social organization and discursive practices," draws on Derrida's analysis to assert that insofar as a

> binary opposition is a "violent hierarchy" where one of the terms forcefully governs another . . . a crucial stage in the deconstruction of binaries involves their inversion, an over-turning. (65)

Thus Dollimore quotes Derrida endorsing "the necessity of the phase of reversal":

> "To neglect this phase," [says Derrida,] "is to forget that the structure of the opposition is one of conflict and subordination and thus to pass too swiftly, without gaining any purchase against the former opposition. . . . [To neglect this phase] deprives one of any way of intervening effectively." (65)

Among the many reversals Aidoo deploys, she sets the terms of her own strategy for reversal very early on in her account of Sissie's travels by reworking a scene from Fanon's fifth chapter in *BSWM* entitled "The Fact of Blackness," which begins: "'Dirty Nigger!' or simply, 'Look, a Negro!'" (111).[12] On her arrival,[13] and while in transit from Frankfurt to the "small [Bavarian] town" where she'll work during her stay in Germany, Sissie wanders through the shops at the railway station; she hears:

> Music. Sounds. Noises.
> So many different noises mixed together.
> Suddenly she realized a woman was telling a young
> girl who must have been her daughter:
>
> **"Ja, das Schwartze Mädchen."**
>
> From the little German that she had been advised to study
> for the trip, she knew that "das Schwartze Mädchen" meant
> "black girl."
> She was somewhat puzzled.
> Black girl? Black girl?
> So she looked around her, really well this time.

> And it hit her. That all that crowd of people going and
> coming in all sorts of directions had the colour of the pickled
> pig parts that used to come from foreign places to the markets
> at home.
> Trotters, pig-tails, pig-ears.

Sissie's ruminations conclude with:

> For the rest of her life, she was to regret this moment
> when she was made to notice differences in human colouring.
> But what she also came to know was that someone some-
> where would always see in any kind of difference, an excuse
> to be mean.
> A way to get land, more land . . .
> Gold and silver mines,
> Oil
> Uranium
> Plutonium . . .
> Jewels . . .
> Houses . . .
> A harsher edge to a voice.
> A sharper ring to commands.
> Power, Child, Power. (12–13)

I have quoted at some length from Aidoo's text because I want to bring attention to its unusual typography, the heady mix of genres and registers through which Aidoo achieves her effects (more of which later). For the moment, I wish to focus on the responses the passage above is likely to elicit and has elicited from its readers, which in turn will allow me to speculate about some of the agendas that inform Aidoo's use of reverse discourse.[14] Shock, discomfiture, embarrassment such as one feels at a joke in deliberately bad taste, a crude remark that one might self-deprecatingly exchange in private, but one that here is articulated in a public space—all these and more the reader is likely to feel at the specification this time of "whiteness" (what Elizabeth Abel calls "that elusive color that seems not to be one" [497]), which is rendered visible, grotesquely so, through its analogy with "pickled pig parts." At first glance, Sissie's analogizing of whiteness with "pickled pig parts" appears harsh, a retaliation in excess of what is a seemingly innocuous remarking of difference—"A black girl?"—through the specification of color.[15] But as Sissie makes clear, "blackness" (and its concomitant "whiteness") are socially constructed categories embedded in a long history of colonial domination in which color came to signify a differentiation on which an entire edifice of racial thinking and colonial adventurism was built. Alluding to the ideological processes through which "whiteness" and "blackness" acquire meaning and evaluative freight enables Aidoo to denaturalize valuations presumed to be naturally given. By transposing through the very coarseness of her analogy the valuations usually associated with "whiteness," Aidoo expects to provoke a shock as great as the one Fanon felt (and Sissie feels when she is described as "a black girl") when "the fact of [his] blackness" was brought home; when he found himself an "object in the midst of other objects" (*BSWM* 109).[16] Similarly, it implies the hope that this shock will be followed by self-reflection and self-criticism regarding normatively held (racialized) beliefs.

In order to grasp one of the significant effects of Aidoo's reverse discourse, it is important to distinguish as well between the implications of what are after all parallel experiences for Fanon and Aidoo's protagonist with respect to coming (in)to a consciousness of their race. Both register this consciousness in the metropole; both, moreover, are informed of this by strangers whom they never directly confront. But Fanon's eyes are turned, once he's labeled "A Negro," inward toward himself and his blackness:

> I occupied space. I moved toward the other . . . hostile but not opaque, transparent, not there, disappeared. Nausea. . . . I subjected myself to an objective examination, I discovered my blackness. (8)

Conversely, Sissie's are directed outward, focusing on the "whiteness" of those she sees "going and coming":

> And it hit her. That all the crowd of people going and coming in all sorts of directions had the colour of pickled pig parts . . . (12)

"[A]shamed of her reaction," Sissie "want[s] to vomit" (12). In a remarkably astute essay, Cindy Bello, a student in my course on "History, Gender, and Nation in Post-Colonial Narratives," glosses this difference between Fanon's and Sissie's reactions in the following way:

> For Fanon, the instance of having his race pointed out to him has the effect of shaming him, triggering a self-consciousness of his body which renders him powerless. In this moment, whiteness becomes transparent, normal, healthy whereas his own blackness has suddenly left him "distorted." Sissie, on the other hand, is made conscious of the whiteness surrounding her. Her shame at this moment is not a shame which stems from her blackness—she is shamed by the disgust which infuses her view of whites. In this instance, her own racial difference is not even an issue, she assumes the gaze of agency.

For Bello, the difference in Fanon's and Sissie's responses is crucially tied to gender. If the educated black man, who is the subject of Fanon's *BSWM,* is a slave to his desire to be(come) white, this desire is mediated by his identification with the dominant that depends in large part on the fact of his maleness. Aidoo's female subject, lacking such a connection with the dominant because of her gender, finds an identification with whiteness difficult, even impossible, and is thereby "freed" of the travails of deracination that so affect the black male subject.

3.

Although one of the more compelling criticisms of the use of inversions, as I have already pointed out, is that by reproducing it preserves the structures within which its object of critique is embedded, its effects are not always significantly different from those that accrue, for example, from the strategy of re-play and repetition I have analyzed in the chapter on Rushdie. Insofar as the use of inversions demonstrates how an evaluation or representation can be mustered in support of an entity's presumed diametrical opposite, this use brings attention to the fact of the evaluation or representation's ideologically determined construction, and thereby critically interrogates, even denudes it of, its self-evidentness and power. Consider as another instance how the inversion of colonial travel to the colonized peripheries reveals the ideology of colonial travel in particular but also the ideology of travels to an "alien" or "foreign" land in general. Although the mythology surrounding travel to foreign lands tends to emphasize encounters with the strange and the different, and through that (or because of it) an expansion in the traveller's consciousness, the traveler usually sees (and records) what she or he expects to see—what, in other words, she or he has been wittingly and unwittingly trained to see.

For example, with regard to French and English colonial travel narratives about journeys to and residence in the Orient, Edward W. Said has shown how "direct observation or circumstantial description of the Orient are fictions presented by the writing on the Orient"; in these travel narratives what we get are "re-presentation[s] of canonical material. . . . [P]ilgrimage is after all a form of copying" (*Orientalism 177). Similarly, in* his comments on Charles Kinsley's *At Last: Christmas in West Indies,* Simon Gikandi remarks on

> the extent to which the narrative of travel derives its authority from pretexts as much as from original observation . . . how the narrative of travel functions as an elaborate reworking of the colonial library, a rediscovery of the already discovered. (97)

Examining "Englishness and the Culture of Travel," Gikandi also locates in travel to the colonial peripheries a means whereby the English travelers constituted and understood their Englishness. Thus, although "travel functions," says Gikandi,

> as a form of metacommentary, allowing the imperial travellers to reflect on, question, demonize, and sometimes assimilate "monuments of other times and places" (88)

it also functions with "Englishness" as "always the stable point of reference" (90). The "journey" to the colonial peripheries, in other words, "is . . . overdetermined by the paradigmatic values that define the domestic epos and ethos" (89).[17]

Sissie's "reverse discourse" about her travel to and within Europe has Africa as its "stable point of reference," with Europe functioning this time around as the other of an African self ("And as they left Africa, there was this other continent" [11]). More specifically, Sissie's travel narrative is framed by her deeply felt urgency to rid Africa of neo-colonialism and build a new Africa that is not independent in name only.[18] To this end, she specifies the instruments of neo-colonialism fairly early in her narrative, *before* she even begins her journey. They continue, however, to be targets of her critique throughout her narrative and virtually overdetermine her response to Europe. Thus, Sissie encounters the Europe she expects to encounter, prepared by "textbook illustrations at home:/Known landscapes/Familiar territories" (37), certainly, but also and perhaps more importantly, by her deep dislike of neo-colonialism through whose perpetuation the power of the West continues unabated.

For Sissie, the instruments of neo-colonialism inevitably include the native ruling elite, who for their personal gain and wealth have mortgaged Africa to western interests. These instruments also include all forms of European aid to its former colonies: academic scholarships, international programs such as the one Sissie travels on, monetary help, food, and so on. All these are the insidious means whereby the continuing dependency of former African colonies is secured in the interests of the West's power.[19] By far the most profound instrument of incorporation of the formerly colonized into (neo-)colonial structures is the group constituted by the "nigger who is a 'moderate'" and his "academic-pseudo-intellectual version . . . who in the face of reality that is more tangible than the walls of the slave forts . . . still talks of universal truth, universal art, universal literature" (6). Composed of men virtually identical to the deracinated Antilleans whose "white masks" Fanon excoriates, this groups includes been-tos, like Sammy (whose "voice, as he spoke of that far-off land, was wet with longing," and for whom "going to Europe was altogether more like a dress rehearsal for a journey to paradise" [9]), and Sissie's lover, who opts to stay on in the West instead of returning home. Ultimately, it is the latter and others like him who come in for the sharpest criticism for electing to stay away from Africa. In fact, it is entirely plausible to read *OSK* as an elaborate and pointed diatribe against what Aidoo views as a willed (and willful) homelessness pursued, in Samantrai's fine characterization, in "the nation-defying identification of immigrants and diaspora-dwellers" (141). Aidoo's acute disparagement of migrants as co-opted or, worse, conscious and unconscious instruments of Western hegemony and power is quite unlike Rushdie's privileging of migrants and their resistant positioning within the metropole.

Paradoxically, however, it is this group, composed of a deracinated African bourgeoisie, that Aidoo's protagonist wishes to call upon to build a free Africa. But in order to participate in this task fully and with integrity, this group, which has assimilated to the extent of making virtually its own colonial ideology and habits of being in their world, must be forcefully persuaded, even coerced, to divest itself of this ideology. Moreover, Aidoo's protagonist's choice of this group, and the dilemma it (or she, for that matter) faces is neither idiosyncratic nor restricted to her. For example, Amilcar Cabral, a participant in, and leader of, the liberation movement against Portuguese colonialism in Guinea-Bissau, while deeply critical of the petit bougeoisie made up of the *assimilados* who had internalized "the colonizer's mentality" ("National Liberation and Culture" 45), nevertheless, acknowledged that "for historical reasons" this was the group best positioned to lead the struggle against colonialism and neocolonialism. "For events have shown," Cabral noted,

> that the only social sector capable of being aware of the reality of imperialist domination and of directing the state apparatus inherited from this domination is the native bourgeoisie. ("Weapon" 108)

More significantly,

> by nature of [this bourgeoisie's] objective and subjective position (higher standard of living than that of the masses, more frequent contact with the agents of colonialism, and hence more chances of being humiliated, higher level of education, political awareness, etc.) it is the stratum which most rapidly becomes aware of the need to free itself from foreign domination. ("Weapon" 108)

It is important to note at this juncture that all the reasons Cabral assigned for targeting the petit bourgeoisie as the primary instruments of decolonization have to to do with this group's positioning *within* the folds of the ideology and structures of colonialism. It is precisely because of this group's insider knowledge of the dictates of colonialism, which proceeds from this group having internalized these dictates at the same time as it's denied the genuine insider's access to colonialism's power—this group, after all, is "not quite, not white" in Homi Bhabha's precise phrase—that it finds itself positioned, at least potentially, as a revolutionary group. In targeting the same group as instruments for freeing Africa from the stranglehold of neo-colonialism, Aidoo, depite her starkly oppositional rhetoric, in effect concedes that resistance is not (or not always) uncontaminated in that it is not (or not always) positioned outside the logic of what it resists.

Of course, neo-colonialism is not identical to colonialism. Entailing as it does an "indirect" form of "imperialist domination . . . by a political power made up mainly or completely of native agents . . . controlled by the ruling class of the dominating country" (Cabral, "Weapon" 100–101), neo-colonialism is, says Gayatri Spivak, "what happened after the beginning of the dismantling of colonial power" ("Neocolonialism" 220). Its effects, she rightly notes, "are subtler . . . much harder to pin" because "you feel it less like you don't feel it—you feel you're independent" (221). If we consider that neo-colonialism gives this petit bourgeoisie relatively greater access to power than what it enjoyed under colonialism; if we consider, moreover, that, like converts who are sometimes more loyal to the articles of their adopted faith than those born in that faith, this group's incorporation into colonial ideology is most likely more thoroughgoing, then we are in a position to appreciate how difficult, indeed, virtually intractable, a task it is to persuade this group out of its attachment to its more or less entrenched Westernization. Our Sister Killjoy's abrasiveness, her self-righteous moralism, and her willingness to shock and provoke, and to call people, ideas, and attitudes polemically to account are to a large degree a function of this near-impossible task she faces.

Employing a mix of genres (narrative, prose-poetry, love letter) and a mix of affective registers (sarcasm, ouright anger, humor, pathos), Sissie's (and Aidoo's) argument against the blandishments of the West (which for Sissie and Aidoo constitute the bases for the been-tos' continued enchantment with and the migrants' continued sojourn in the West) proceeds through a series of brutal debunkings, assertions of the West's malevolence and bad faith where benevolence and good faith had been uncritically granted or presumed by the been-tos and the migrants:

> Oppressed multitudes from the provinces rush to the imperial seat because that is where they know all salvation comes from. But as other imperial subjects in other times and other places have discovered, for the slave, there is nothing at the center but worse slavery. (87–88)

Ranged against the voluntary (or enforced) deracination of the Westernized migrants and been-tos, Sissie's (and Aidoo's) argument is, virtually by definition, a closed (pro-return-to-Africa, anti-migration-to-the-West) argument, which exists in an uneasy, even contradictory relationship with *OSK*'s formal "openness" (signaled, for example, by blank pages with only one word inscribed, which almost invite the reader to fill in the blank spaces) and fragmentation, when, for example, prose-poetry disrupts the linear logic of Sissie's narrative. However, this closure, I've been suggesting, is to some extent a function of whom she's addressing. It is also a function of the particular strategies or procedures of the discourses she is reversing.

4.

If one way we can account for the significance of nationalism for Aidoo is by linking it with anti-colonial nationalism, in which context it often represents an emancipatory, counterhegemonic stance and consciousness in resistance to colonial power—a stance and consciousness that continues to be relevant, even necessary, in neo-colonial Africa, according to Aidoo[20]—there seems to be no way, at first glance at least, to account for Aidoo's deployment of nationalism as emancipatory given that strong, independent-minded "women [are her] central characters," indeed, "the center of [her] universe."[21] For as recent feminist interrogations of various nationalisms have revealed, the particular power that mobilizes nationalisms (and the ideologies and discourses that underpin and articulate them) is male power. Nationalisms most often focus on, and themselves issue from, what is constituted as a masculine, public sphere, which not only construes women as having an indirect relation via men (or in the interests of male self-definitions) to the nation, but also relegates women's specific interests and material realities to a privatized, domestic sphere.[22] As Anne McClintock notes:

> Women are typically constructed as symbolic bearers of the nation, but are denied any direct access to national agency. As Elleke Boehmer notes . . . the male role in the nationalist scenario is typically "metonymic"; that is, men are contiguous with each other, and with the national whole. Women, by contrast, appear "in a metaphoric or symbolic role." ("No Longer" 261)

Nationalisms, in other words, by marginalizing or short-circuiting female agency, have seldom if ever proven emancipatory for women.

Of course, one could at this juncture simply agree with Brenda Cooper's assertion that Aidoo and her protagonist assign priority to nationalist identification (particularly "being black") over gender (in this instance, "being female"). There is some textual support for this position. For example, in all her encounters and (brief) relationships with well-meaning white women, Sissie notes how "colour" and a history of colonization inevitably separate her from establishing any meaningful bonds with these women. I think, however, that Aidoo has something different in mind. Aidoo admits that "nationalism *has* been compromised by its marginalization of women." But she also views the "marginalization of women" as endemic to "all areas of human endeavor. I mean women have been written and talked out of all the principal areas of discourse, of activity, of thought" ("An Interview" 124). Her last statement provides an important clue with regard to her deployment of nationalism almost against its grain—with, that is, its usual participants reversed.

For what deserves our attention is not just that Aidoo deploys nationalism as a significant embodiment of counterhegemonic ideology, discourse, and praxis, but also, and more importantly, that it is her female protagonists who are its most vociferous and effective proponents in the public arena. I believe that this bespeaks Aidoo's attempt to insert (*write*) women into precisely those spaces and roles to which they have hitherto been denied full and meaningful access. Thus, Aidoo juxtaposes our sister's insistent appeals for, and her own return to, Africa with the "brothers'" sullen recalcitrance in the face of her appeals, and their elaborate and, by Sissie's reckoning, spurious justifications for staying on in the West: "'Listen, listen, listen,' [says one,] 'I have made our people proud . . . [g]aining recognition as one of the world's ten experts on gastric disorders'"; and another: "'But you can see how by remaining here someone like me serves a very useful purpose in educating them to recognize our worth . . . ?'" (127, 129). For Sissie, these are the justifications of a "superior monkey [who] has got his private white audience for whom he performs his tricks" (130). The "brothers" who migrate to the West "function," as Ranu Samantrai so aptly puts it, "as the foils against which Sissie works out her allegiance [to Africa]. Their migrancy becomes emblematic of the wrong response to the seductive call of the overdeveloped west" (152). By the same token, Sissie's refusal to be seduced by the blandishments of the West constitutes the right response.

The binary thinking that organizes Sissie's view of resistance to colonial and neo-colonial ideologies is therefore also much in evidence in Aidoo's attempts to reverse the roles nationalisms usually assign women and men. In the short story that opens her collection *No Sweetness Here,* "Everything Counts," and in *OSK,* not only do Aidoo's female protagonists occupy the center, as it were, and the men the margins, but these women also function as the carriers and voices of a genuine nationalism while the men pursue only a spurious version.

"Everthing Counts" represents the debilitation and dependency fostered by neo-colonialism through arguments about the inappropriate politics of wearing wigs—not an"Afro-wig," says Aidoo, but "straight hair" through which Africans seek to "remake" themselves "in the image of the colonizer" ("An Interview" 132). In this story, the female protagonist, also named Sissie, who has been educated abroad and has married a fellow African abroad because she does not want to deal with family pressures and politics back home, returns home to Africa and to the debate about whether wearing wigs is politically correct. Initially, she rejects what her "brothers" settled in "Europe, America, or some place else" (7) dictate to the young women back home that "wig-wearing is wrong" because it "is totally foreign. Unhealthy" (6). "But what does wearing wigs have to do with revolution?" she queries in exasperation (2). However, after observing how the judges at a beauty contest inevitably select the light-skinned, straight-haired contestants, thereby signaling their continuing subjugation, in this instance, to Western conceptions of beauty, Sissie concedes that the "brothers" were "right," that "wig-wearing is wrong" because it incorporates Africans further into the folds of (neo-)colonial dependency. It is significant, however, that although the story affirms Sissie's "epiphany," it dismisses as mere rhetorical posturing the "brothers'" identical claim about "wig-wearing." Why? Because, as Aidoo put it in her interview with me, they "are not even there [i.e., in Africa]"; they "don't even know what [they] are saying . . . couldn't even have perceived what [Sissie has] perceived" ("An Interview" 132–133). Thus what distinguishes a genuine from a spurious nationalism is not the contents of a given belief or principle (after all, both Sissie and the "brothers" come to agree that "wig-wearing is wrong") but rather who espouses the belief or principle and from what location. Having returned to Africa, and so, by implication, having participated in its material realities, Sissie, Aidoo's story suggests, has earned the right to her "epiphany"; conversely, the "brothers" away in "Europe, America, or some place else" have not.

Aidoo's stories in *No Sweetness Here* every once in a while complicate the binary through which women emerge as genuinely resistant subjects and men as their obverse—lacking a comparable integrity, compromised by their brute self-interest.[23] Nevertheless, several of her female (and emphatically not male) characters come through as the most clear-sighted and least likely to compromise their insight into the inequities and oppressions fostered by neo-colonial arrangements. In a narrative aptly titled "For Whom Things Did Not Change," Zirigu, one of the central male characters, who looks after the guest house to which the "big men" come, wonders what "independence" means.[24] "Independence" has a hollow ring for him because the symbols of his access to some measure of human dignity and worth promised by independence—electric lights and a water closet in the lavatory instead of an "old pail"—are denied him, the "old pail" being replaced by a "new one" with the arrival of independence (28). But despite his recognition of the failures of independence, despite, moreover, his recognition that these failures can be laid at the door of the "big men," he accedes to his lot in life when he grants the "big men" special status:

> "When a man is your lord, he is your lord. And he behaves like your lord. How else should he behave? And who are we to say that new lords must not do what the old ones did? When the white men were here did they not do the same? Sleep with very little girls, oh such little girls?" (11)

If neo-colonial rule, following so seamlessly on the heels of colonial rule, can be explained by alluding to its beneficiaries, the ruling elite and native bourgeoisie who moved into the spaces only recently vacated by their colonial counterparts, it is difficult to understand the consent (or passive acceptance) of neo-colonial arrangements by someone like Zirigu who has been left out of their reckoning altogether. Aidoo's narrative suggests that Zirigu's acceptance of dominant ideology may well be a function of his being male. For his wife has no compunctions whatsoever about articulating her anger and dissent. "'O Allah, what times we live in,'" she exclaims, "'What rulers we have. How can they behave in this way who are our lords?'" "'It is good I never had a daughter,' she adds 'Because if I had a daughter, and I knew a big man was doing unholy things with her, then with a matchet in my own hand, I would have cut him to pieces myself'" (11). Predictably, Zirigu's response to his wife's verbal violence is to evoke quietism and religion: "'O, Jesu preserve my soul. O Jesu! Setu what kind of talk is this? You must pray more often than everybody on Friday for these foul words'" (11). Thus Zirigu, left out of the benefits that accrue to domination, nevertheless remains trapped in its ideological agendas, granting them his mournful consent.

5.

If one were to specify what for Aidoo, so far as is evident in her work, constitutes the essential attributes likely to produce a genuinely resistant subject, these attributes would be African, black, female, with a staunchly anti-Western stance and sensibility, where Western is conceived as coterminous with imperialism. Such an stark inventory, while more or less on the mark about "our sister" Sissie, is, some would argue, a little too flat and uncomplicated to be able to account for the more nuanced characters and the dilemmas and challenges they confront in some of Aidoo's short stories in *No Sweetness Here,* which, being in the realist vein, aspire to some degree of psychological density. Even in these stories, however, these attributes continue to be fundamental determinants of genuine resistance, though it is worth reiterating that resistance in Aidoo's work is always framed by the context of neo-colonialism and thus by her need to rid Africa of its continuing stranglehold.

Aidoo's view of resistance, then, framed by a specific set of circumstances and imperatives, relies on essentialist categories—carefully policed inclusions and exclusions as they pertain to ideological positions and identi(ty)fications. The limitations of the procedures she deploys have been rehearsed quite extensively, as have their virtues or, better, their necessity, even inevitability. The most oft-repeated objections against the use of essentialist categories and thinking are rooted in a recognition of, indeed insistence on, the inherent heterogeneity of things, ideas, peoples, positions, and so on. These objections often remark on the multi-accentual, ambiguous, internally fissured nature of even those entities, ideas, values, and beliefs that appear unified and cohesive. Remarking on a given essentialism's deliberate suppression of differences, critics also comment on how any essentialism's deployment disregards or elides historical thinking, imposes false closures, and tends toward absolutist positions that can be dangerous and discriminatory—at least as dangerous and discriminatory as those they oppose.

Among defenses of essentialist thinking and categories that do not evade or disregard these dangers, perhaps one of the more significant ones is articulated by Gayatri Chakravorty Spivak:

> it is not possible, within discourse, to escape essentializing somewhere. The moment of essentialism or essentialization is irreducible. In deconstructive critical practice, you have to be aware that you are going to essentialize anyway. So then strategically you can look at essentialisms, not as descriptions of the ways things are, but as something one must adopt to produce a critique of anything. (*Post-Colonial Critic* 51)

Or to put it another way, essentialisms and essentializations possess, to borrow Alcida Ramos's terms, "heuristic" and not "empirical" value; they are "operational devices," not "ontological realit[ies]" (230). Elsewhere Spivak speaks of the Subaltern Studies collective's "strategic use of positivist essentialism in a *scrupulously visible political interest*" ("Deconstructing Historiography" 342).

We can, as I have tried to demonstrate through my analysis of Aidoo's work, invoke *strategy* for Aidoo's use of essentialist categories, which can and must be read in the context of her anger about and desire to wrest

Africa from its downward spiral into further neo-colonial dependency. "'African identity,'" as Benita Parry reminds us in her examination of Fanon and Césaire, and their connections with the Négritude movement, "is the product of refusing Europe's gaze and returning its own anti-colonial look" ("Resistance Theories" 177). Emphasizing strategy, and thus treating essentialist attributes like African, black, and female, so salient to Aidoo's elaboration of anti-imperialist resistance, as in fact possessing "heuristic" value, we can view Aidoo's procedures as politically progressive, even necessary for achieving the conditions for a free Africa. The difficulty (to which critiques and endorsements of essentialism bear vivid, albeit implicit, witness), of course, is that the line separating "an intellectual apprehension of blackness as a construct" from "a visceral attachment to the powerful fiction of black identity," as Parry notes, is often hard to maintain. Futhermore, to recognize a representation or stereotype as a politically interested figuration does not denude that representation or stereotype of its material effects—whether negative or positive. Any number of examples from colonialist and anti-colonialist, racist and anti-racist discourses will testify to this. It is sometimes impossible not to read Aidoo's deployment of these essentialisms as ambiguously situated between an awareness of these as necessary but invented constructs and her "visceral attachment" to their truth or reality. Recognition of this ambiguity should provoke recognition about the limits that inevitably accrue to forms of enablement or, in this instance, to forms of resistance. To adapt a comment from Spivak, we continue, however, to use them, because without them we "cannot live on, take chances" ("In a Word" 4).

Chapter 4

"Retracing the African Part of Ourselves": Blackness as Revolutionary Consciousness and Identity in the Work of Michelle Cliff

> *If we ally ourselves in the struggle with the oppressed (and especially if we are working in the context of privilege), how will our fictions, our theory, name this solidarity?*
>
> —bell hooks

> *History is a moral force from which societies drink deep in order to reconstruct their consciousness and their identity. . . . To deprive a society of its history, to forbid it the production and knowledge of its history, is to assassinate its genius, to render it foreign unto itself. The falsification of history is a crime against its people, the search for historical truth . . . is a mission entrusted to the "organic" intellectuals.*
>
> —Bogumil Jewsiewicki

I.

Tracking her "journey into speech,"[1] which forms the prefatory matter of her collection of prose and poetry, *The Land of Look Behind,* Michelle Cliff remarks on Ama Ata Aidoo's *Our Sister Killjoy* as an enabling catalyst: "After reading *Our Sister Killjoy,* something was set loose in me" (16). Remarking on how she "was especially drawn by the way in which Aidoo expresses her rage against colonialism," so that Aidoo's "prose breaks apart sharply into staccato poetry—direct, short, and brilliantly bitter," Cliff aspires to a similar rage and directness in her work:

> I wanted that kind of directness in my writing as I came into closer contact with my rage, and a realization that rage could fuel and shape my work. . . . After reading Aidoo, I knew I wanted to tell exactly how things were, what had been done, to us and by us, without muddying the issue with conventional beauty. . . . I let myself go, any thought of approval for my words vanished . . . I . . . tore into the indoctrination of the colonizer, surprised myself with the violence of my words. (15–16)

Indeed there are powerful bonds that tie the work of these two writers together, not the least of these being their focus on the damage that colonialism, and now neo-colonialism, has wrought in their homelands, and their attempt to make the colonized psyche (their own and their compatriots') whole by advocating a "return to [their] native land[s]." They envision (one implicitly and the other explicitly) these lands as female, but by making literal what is in most patriarchal, nationalist formulations only a symbolic or metaphorical designation. By doing so, both Aidoo and Cliff subvert dominant definitions of nationalism that situate men in the public arena—the arena conceived as vital to the workings of the nation state—and women in the private arena. Furthermore, both Aidoo and Cliff link the idea of a "return to [the] native land" with a commitment to and an alliance with Blackness (which Cliff also defines as "the African part of ourselves [i.e., herself and her fellow-Jamaicans]").[2] The idea of "return" linked with allegiance to Blackness in turn embodies for both writers a form of revolutionary praxis and identity.

However, whereas Aidoo's advocacy of and allegiance to Blackness as an exemplary form of resistance against colonialism and neo-colonialism is premised on the continuity between her identity and identification or allegiances, Michelle Cliff's is not and perhaps cannot be. Born in 1946 in Jamaica, Cliff's family moved to New York when she was very young. Educated in public schools in New York, she earned her bachelors degree in European history from Wagner College. After a year in publishing (she worked for W. W. Norton), Cliff left for London to pursue her master's degree in philosophy. "Radically deterritorialized" (195), as Caren Kaplan puts it, from the Caribbean and its culture(s), she is "deterritorialized" also because as a "Jamaica white," an identity Belinda Edmondson glosses as "a Jamaican of mostly white ancestry" (181),[3] Cliff is presumed not to have unmediated access to Blackness.[4] Thus the Blackness Cliff, or her autobiographically informed protagonist in *Abeng* and *No Telephone to Heaven,* Clare Savage, claims is an identification she works at and earns the right to via a series of choices she makes, and by purposefully affiliating herself with certain positions, peoples, and modes of be-ing in the world. To be sure these affiliations and identifications also include those that she perceives as having an unmediated access to revolutionary consciousness and identity that for her Blackness represents. But in elaborating upon and (implicitly) discriminating between them by juxtaposing them with those that have a self-consciously mediated access to Blackness, Cliff makes visible a paradox—regarding identity as both given and constructed—that underpins Aidoo's work as well, except that Aidoo does not explicitly unpack it in her work, leaving it instead to her readers to do so. In this regard, then, Cliff's view of what makes a subject, position, or action resistant—what constitutes resistance—serves as a useful albeit unintended gloss on Aidoo's, inviting their readers to discriminate between and evaluate their formulations about identity and identifications. Furthermore, affiliating with Blackness as opposed to already "possessing" it, as, for example Aidoo is presumed to, commits Cliff and Clare to a teleology of sorts whereby they are presented as acquiring or coming to "revolutionary consciousness" (the phrase is Françoise Lionnet's) over time and through the density of identifications also accumulated and made sense of over time. Cliff, in other words, delineates the process leading to, as well as specifying what constitutes, resistance and/or resistant subject positions.

In this chapter, I examine this process and analyze what Cliff defines as resistance and/or resistant subject positions. I do so by focusing on several of her fictional and nonfictional works: *Abeng* (1984), *No Telephone to Heaven* (1987), selections from *The Land of Look Behind* (1985), and finally a passing glance at Cliff's foray, in *Free Enterprise* (1993), into a forgotten or ignored part of American history—Mary Ellen Pleasant's contribution to the abolitionist cause. I am especially interested in exploring the modalities through which Cliff seeks to retrace "the African part of ourselves [i.e., herself and her fellow-Jamaicans]" in order to present Blackness as a significant resistant positionality. These modalities include: (1) excavating a hitherto suppressed or disregarded archive of knowledge about Jamaica: the hidden or ignored history of resistance against slavery and colonialism;[5] (2) aligning herself and Clare with people, ideas, and positions that exemplify struggles against various forms of subjugation, through which Cliff and Clare then define their own communities in struggle, "name [their] solidarity," as hooks puts it, through and because of a shared oppression.

2.

History as subject, idea, (narrative) genre, and strategy is a central preoccupation of Cliff's work. Virtually her entire oeuvre testifies to this. However, Cliff is invested, like the writers we have examined thus far, in a particular kind of history—an alternative history—which is distinct from the one she calls "official," and which she polemically dismisses in the following remarks:

> The official [or hegemonic] version has been printed, bound, and gagged, resides in schools, libraries, the majority unconscious. Serves the common good. Does not cause trouble. Walks across tapestries, the television screens. Does not give aid and comfort to the enemy. Is the stuff of convocations, colloquia; is substantiated like the Host—in dissertations. . . . The official version entertains. Illumines the Great White Way. (*Free Enterprise* 16–17)

By definition, an alternative history is not (and cannot be) any of the above. Instead, as Cliff puts it in *The Land of Look Behind,* in the context of colonial and neo-colonial Jamaica, an alternative history is one that needs to be retrieved:

> a history sunk under the sea, or scattered as potash in cane-fields, or gone to bush, or trapped in a class system noted for its rigidity and absolute dependence on color stratification. (14)

Whether a history is remembered or forgotten or buried is a function most often, as the remarks above suggest, of deliberately undertaken censoring mechanisms, and thus one tied to specific agendas and interests. What are some of the underground histories that Cliff wishes to resurrect and make a part of her people's (and her readers') overt consciousness?

Usually the histories that are suppressed or ignored are those that do not reflect or redound favorably on dominant culture and the ruling classes—their beliefs and values. In the context of colonial and neo-colonial Jamaica, according to Cliff, these suppressed or ignored histories include accounts of struggles against colonialism;[6] of slave insurrections; of rebellions, moreover, in which women played prominent roles, but whose very presence is ignored even when resistance is acknowledged.[7] In her interview with Judith Raiskin, noting how "most of [her] work has to do with revising: revising the written record, what passes as the official history, and inserting those lives that have been left out" (71), Cliff notes how her alternative history then attempts to set the record right by reconstructing and making visible these hitherto hidden histories of resistance and inserting within them the crucial part played by women.

In *Abeng,* for example, Cliff intersperses Clare's coming-of-age story with an account of the "armed, sustained guerilla warfare" (20) of the Windward Maroons, usually left out of or downplayed in dominant versions of Jamaican history.[8] These Maroons, we are told, "held out against the forces of white men longer than any rebel troops" (14). The word "Maroon," Cliff remarks, comes from "*cimarrón:* unruly, runaway . . . the word connoted fierce, wild, unbroken" (20). The pride of place in these Maroon wars against plantocratic culture Cliff assigns to the person she names "the magician of this revolution," Nanny—an extraordinary and powerful woman, a skillful military tactician and leader whose power and skill Cliff traces back to Nanny's ongoing organic connection with Africa. Thus Nanny "was from the empire of Ashanti and carried the secrets of her magic into slavery"; she could spin "Akan chants" and "[r]emember the battle formations of the Dahomey Amazons" (14; 21). Nanny appears an unexceptional "small and old Black woman," but one who wears a "necklace fashioned from the teeth of white men" (21). Defiant to the end, she does not die a natural death, nor even one in battle, but is betrayed by a fellow Maroon turned collaborator of the colonizers, Cudjoe, leader of the Leeward Maroons.

In this alternative history, privileging resistance by the enslaved and women, Cliff shows how Nanny's courage and power are replicated in Mma Alli, a slave woman well-versed in *obeah* and magic, who "had never lain with a man. . . . She loved only women that way, but was a true sister to the men—the Black men: her brothers" (33). And in Inez, Judge Savage's forcibly acquired "bronze" mistress, who, despite her belief that children are sacred, goes to Mma Alli "to get rid of the mixed-up child she carried" because "a baby conceived in *buckra* rape would have no soul" (35). [9] What these women have in common with the other rebellious Maroons is their love for their own people—Black people—and for Blackness, and their refusal to be subjugated by the colonizers; what these women have in common with each other is their love as women for other women.

Among the hidden or forgotten facts that Cliff's alternative account in *Abeng* resurrects are the atrocities committed by the plantation owners, managers, and overseers against the slave populations. Remarking on the exceptional brutality of the slave society in Jamaica,[10] Cliff notes how to prevent slaves from eating the "sugarcane they cut" the "slaves worked with their faces in masks of tin"; how with "few white women on the island during slavery," slave women were "violated again and again by the very men who whipped them." Though these women would have been raped "with or without the presence of white women," Cliff concludes that "in Jamaica there was no pretense of civility—it was all out in the open" (18–19). A searing indictment of colonial, plantocratic culture, her catalogue of horrors is meant to suggest another (invisible to dominant culture and its discursive apparatus) source of resistance as well. Operating, like Stella Dadzie, on the principle that "the degree of oppression in Jamaica" bore "a quantifiable relationship to the intensity of the slave resistance on the island" (Dadzie 26), Cliff juxtaposes these passages on the brutalities directed against the slaves with passages recounting rebellions and acts of defiance committed by them. Resistance, in other words, is viewed as existing in dialectical interplay with oppression. This is why although Cliff retrieves these suppressed or forgotten histories in general to disseminate an alternative account that challenges or corrects the dominant one,[11] she is interested in particular in making them known to those oppressed people and subaltern groups for whom they matter (or should matter) but who as yet have no knowledge of them. One such group is comprised of the congregation at the Tabernacle, composed mostly of poor, Black women who have suffered the most and benefited the least from colonial and neocolonial arrangements that have dominated Jamaica and continue to do so. Cliff's alternative history documenting oppression and resistance is in fact framed by a specific reference to the congregation at the Tabernacle, who, says Cliff, "could trace their bloodlines back to a past of slavery," although it "was not something they talked . . . or knew much about" (18). Cliff's hope is that this knowledge, thus far unavailable or not known and now made available and known, will empower this group of people and others like them to act and thus to come into possession of their hitherto denied or suspended (revolutionary) agency. In an interview with Opal Palmer Adisa, for example, Cliff explicitly links the ability to resist—indeed, the belief that one can resist—with one's knowledge about prior resistances undertaken by one's people in similar circumstances: "It seems to me," she avers, "if one does not know one's people have resisted, then it makes resistance difficult" (280).

History, according to a well-known truism, belongs to the victors, is written by and valorizes them. In the context of colonialism, it was not just the histories of resistance that were suppressed, but the histories in general of the colonized populations that were rendered "subterranean"[12] or simply disappeared. Bogumil Jewsiewicki, remarking on the integral connections that obtain between a people's history and their identity, remarks on the alienation which follows for a people from being deprived of their history ("To deprive a society of its history . . . is to render it foreign unto itself" [96–97]). This alienation is endemic to a colonized population, whose histories, forced underground, are replaced by the history of, and written by, the colonizer. This history predictably redounds favorably on the colonizer and diminishes or disregards the colonized. Thus, in *Abeng,* even when in her colonial school Clare is "taught . . . that Jamaica had been a slave society," this lesson is accompanied by a prompt disclaimer that "England was the first country to free its slaves"; when she is told about "a freedmen's uprising at Morant Bay in 1865, led by Paul Bogle," that "rebellion" is dismissed as "unwarranted" and inconsequential. Hence the history of the colonized in dominant colonial accounts appears (and is) "slight[(ed)] compared to the history of Empire. The politics of freedmen paled beside the politics of commonwealth" (30). No wonder, then, a "return to [and of one's] history"[13] is a vital step in the processes leading to freedom from colonial control. No wonder, as well, recovering a colonized people's "subterranean" histories is deemed to have a "practical function, a revolutionary function that aims to give the consciousness of their continuity in time to [a] people."[14]

3.

Among the long-term (some would say still ongoing) damages wrought by colonialism is the internalization on the part of the (formerly) colonized populations of the colonizer's belief in his own superiority and the colonized's inferiority, rendered visible via color, "white" over "Black"—the "epidermalization," in Fanon's astute description of what is represented as positive or negative, superior or inferior (*Black Skin, White Masks* 11). According to Cliff, even in post-independence Jamaica—with its hybrid populations composed of "Indian, African, European" (in her interview with Meryl Schwartz 595); or "black, white, and mulatto" (Edmondson 182)—whiteness (as both literal and metaphorical referent and condition, as I explain below) continues to dominate, and be valued more, over Blackness (also literal and metaphorical as I explain below), whose histories, because of their connections with a despised or disparaged identity, continue to be disregarded or suppressed.[15] However, given, as Cliff notes in *Abeng,* that "the population of the island was primarily Black . . . with gradations of shading reaching into the top strata of society," and though Africans were mixed with other racial and ethnic groups, "the working class, the class of the poor people was Black" (5), the identity and histories most in need of restoration are those viewed as Black. In fact, Cliff identifies the project of rendering a Jamaican/Caribbean identity whole with "retracing the African part of ourselves": "To write as a complete Caribbean woman, or man," she says in *The Land of Look Behind,* "demands of us retracing the African part of ourselves" (14). This demand, I believe, significantly (albeit not entirely) determines the histories (and identity) Cliff does recuperate, in which Blackness is a primary factor and referent.

But there is to my mind an even more significant determinant of Cliff's focus on and near exclusive interest in Blackness. As I suggest in my opening comments to this chapter, for Cliff, Blackness is co-extensive with a revolutionary consciousness and identity, not the least because it is indelibly associated in her mind with women and with femaleness, which are for her unarguably revolutionary subject positions.[16] In addition, clustering around Blackness, in Cliff's work, are a set of associations—subject positions representing the working class, the poor, the economically and politically dispossessed—that she also views as potentially revolutionary. In Cliff's estimation, then, identification with and allegiance to Blackness is fundamental for forging a resistant consciousness and identity.

Both *Abeng* and *No Telephone to Heaven* (hereafter *NTTH*) explore this identification and the processes through which it is achieved via the coming-of-age narrative of Clare Savage, "a light-skinned daughter of the middle class," the description on the back cover of *Abeng* informs us, "who grows up among the contradictions of class and color, blood and history, in a colonized country." It is neither hyperbolic nor reductive to view the trajectory of Clare's coming-to-mature-self-consciousness of her values and beliefs as a trajectory toward (and an affirmation of) Blackness, or, more accurately, a trajectory toward her full identification with and commitment to her "native land," Jamaica, which she identifies as primarily African and female.[17] But before she can get there literally and figuratively, she must negotiate other spaces and identities, alliances and allegiances, each with its powerful hold on her, and many of them deeply hostile to everything that Blackness represents. (Black and white, it's worth emphasizing here, while referring to the color of one's skin, refer as well to class affiliations, social position, demeanor, a way of inhabiting the world premised on an assumption, or the lack thereof, of access—indeed, entitlement—to power.[18])

As the light-skinned, first-born of Boy Savage and Kitty Freeman, Clare is claimed by her father as his, with the tacit, sometimes explicit support of her mother, who claims Clare's dark-skinned sister Jenny as her own: "Clare," we are told, "had been handed over to Boy the day she was born" (*Abeng* 128); indeed, Boy insists: "You are my daughter. You're white" (*Abeng* 73). This claim is neither self-evident to Clare nor does she accede to it uncritically. Drawn to her mother who "cherishe[s] darkness" (*Abeng* 127), Clare is drawn as well to the spaces associated with her mother. From the outset, Boy and Kitty are described as representing mutually exclusive oppositions between white and Black, town and country, wealth and poverty: Town and country, says the narrator, "reflected the separate needs and desires of the two parents . . . Kitty came alive only in the bush, while Boy armed himself against it" (*Abeng* 49). Whereas Boy comes from a family "well-known . . . for its former wealth," (*Abeng* 22), Kitty's family, the Freemans, were poor" (*Abeng* 135).

Boy willingly, even enthusiastically, migrates to the United States, but Jamaica remains Kitty's "point of reference—the place which explained the world to her" (*NTTH* 66). Possessing different pasts, they perceive their positions relative to the past of slavery in diametrically opposed fashion: for Clare's father, it is "Cambridge University and sugar plantations and a lost fortune." But for Kitty that past recalls "What had been done to her people when they were slaves" (*Abeng* 128).

Even the act of naming their first born, Clare, is fraught with opposing significance and meaning for each: Kitty lets Boy believe that he is naming Clare "after the college [Clare's] grandfather attended at Cambridge University," whereas Kitty in reality has named her after "Clary, the simple-minded dark girl who fought for her" (*Abeng* 141). But having decided that Clare belongs to and with Boy,[19] Kitty does not inform her about Clary; thus "Clare never knew whom she was called after—whom she honored" (*Abeng* 141). When Clare casts her lot with Blackness, one could say that she instinctively honors this allegiance she is not even aware of, for by then, Kitty is long gone.

Despite an acknowledgment of positions located between (and within) Black and white that complicate notions of a monolithic Blackness or whiteness (both Clare and Cliff, after all, are "daughter[s] of landowners, native born, slaves, emigres, Carib, Ashanti, English" [*NTTH* 5]), Cliff presents the choice facing Clare, if she is to resolve the contradictions represented by her heritage, as a choice between either Black or white, emphatically not a choice that invites or urges her to incorporate these conflicting sides within her or to inhabit a psychic terrain defined by a sort of in-betweenness straddling both.[20] In *Abeng,* for example, Cliff, noting Clare's fascination with Pip, the hero of Charles Dickens's *Great Expectations,* and his fate, with which she compares her own, observes how the "idea of a benefactor capture[s]" Clare as she "wonder[s] who her benefactor would be": Magwitch or Miss Havisham, the "Black or the White. Both perhaps . . . She was both dark and light. Pale and deeply colored. To whom would she turn if she needed assistance? . . . Her [dark] mother or her [white] father [?] . . . Would her alliances shift at any given time [?] . . . The Black or the White? A choice was expected of her" (36–37). In *NTTH,* Harry/Harriet reiterates and reaffirms this position when, in response to Clare's statement that the affinity between the two of them is a function of their being "neither one thing nor the other," s/he warns that

> the time will come for both of us to choose. For we will be called upon to make a choice. Cast our lot. Cyaan live split. Not in this world. (131)

Nevertheless, complexities such as those that inhere in, and accrue from, Clare's/Cliff's[21] hybrid ancestry and location are very much at issue in the manner in which she conceives of resistance. With the instruments of empire—including its epistemologies, value system, modes of discrimination and control—still very much in evidence in post-independence Jamaica and an integral dimension, more significantly, of Clare/Cliff's social class and psyche and thus difficult to simply bypass or reject, Clare/Cliff must work with and through these "contaminated" instruments before she can emerge on the other side into her revolutionary identity and consciousness.[22] Cliff's narratives signal that she is fully cognizant of this dilemma; they also demonstrate how she works with and through the instruments bequeathed by empire in order to remake them so that they are redirected into becoming potential instruments for resistance against empire.

For example, the opening pages of *NTTH* make much of the khaki in which the guerillas Clare has joined on her return to Jamaica are clad. "Why make so much of it?" Cliff queries. "It's simple," she responds: khaki is a vestige of empire ("How many stiff bolts of the stuff had been spun to outfit the empire?"[6]). Now, however, its "purpose ha[d] turned"; it was being used to outfit guerillas "evoking the name of Nanny, in whose memory they [were] engaged [in their insurgent activity]" (5). Once again, the "goal" in wearing khaki was, like empire's, to achieve "alikeness," but this time in the interests of forging an alliance and a unity—necessary for carrying out their insurgent work—among a group of people widely separated by skin color and social class, but also by "places traveled to and from, events experienced, things understood, food taken into their bodies . . . books read, music heard" (4). Furthermore, along with another instrument of domination, American army fatigues, khaki served as camouflage. (Camouflage, too, is double-edged—on the one hand used for presumably revolutionary ends to protect the guerillas, on the other used for presumably reactionary ends by Boy, who advises Clare of its uses for blending into dominant American culture when they migrate to the mainland [*NTTH* 100].)

The *abeng* (from which Cliff's first novel about Clare and her childhood and transition into adolescence takes its name) is also, like khaki, an instrument that can be deployed for domination or resistance, depending on the uses to which it is put and by whom. *Abeng* is an "African word meaning conch shell," and "blowing" it "called the slaves to the canefields." But it was also the means by which the Maroon armies communicated with each other. Furthermore as an image—a metaphorical referent—for Clare, the *abeng* signals Cliff's belief that, as a "crossroads character,"[23] Clare (/Cliff) is potentially open to being an instrument of domination or resistance, depending on the uses to which she puts herself and the choices she makes.

4.

"At the end of *No Telephone to Heaven,*" says Michelle Cliff, Clare Savage

> has cast her lot, quietly, tentatively, but definitely. She ends her life burned into the landscape of Jamaica, literally, as one of a small band of guerillas en-

> gaged in a symbolic act of revolution. . . . In her death, she has complete identification with her home-land; soon enough she will be indistinguishable from the ground. Her bones will turn to potash as did her ancestors' bones. ("Clare Savage as Crossroads Character" 265)

This passage, which asserts Clare's organic connection with her homeland, with Clare "literally" inscribed onto and "indistinguishable" from "the landscape of Jamaica," also implicitly suggests that this connection in turn connects Clare with Blackness as well. It achieves the latter connection by, for example: (1) repeating virtually verbatim an image through which Cliff represents the "subterranean" history of Jamaica ("a history . . . scattered as potash in the canefields") that, as I note earlier, has Blackness as its primary referent; (2) recalling Cliff's account of Judge Savage's recaptured runaway slaves, who, subjected to inhuman torture—lashed, hanged, quartered, re-hanged—in their death make his plantation on Runaway Bay "rich and green" (*Abeng* 40): "The bones fell to the ground where they melded with the earth, fertilizing the cane with potash" (*Abeng* 30). Thus the ancestors Clare comes to resemble most closely after her death are the African men and women who labored on her great-great-grandfather's plantation, even though she begins her life more clearly identified as white and claimed by her father as white.

But where Clare's "complete identification" with her homeland is explicitly stated, albeit in an essay by Cliff and not in *NTTH* itself, the suggestion that this is similar to an identification with Blackness is more a matter of indirection—implied rather than asserted, and dependent on the reader making connections between passages that appear at some distance from one another and/or in different texts. The reader of Cliff's work is called upon to make other kinds of connections as well, with respect again to her unpacking of Clare's/Cliff's politicization, whose culmination is presented as her coming (in)to a fully informed and self-conscious understanding of the revolutionary properties of Blackness.

Thomas Cartelli rightly remarks on the "epigraphic authority" *NTTH* claims "for other postcolonial writers and for the traces of postcolonial cultures." Cliff's "chapter headings," he adds, "are studded with quotations from fellow West Indians like Derek Walcott and Aimé Césaire, from Yoruba hymns, and Jamaican proverbs. And her text often alludes to the work of Jean Rhys, C. L. R. James, and the dub poet Linton Kwesi Johnson among others" (90). Indeed, Cliff's other works are also "studded" with references to writers and fictional characters in their work. But this invocation of writers, fictional characters in their work, and cultural forms also serves another, to my mind, more important function: Paralleling Clare's/Cliff's own location as colonial, hybrid subject and the dilemmas or difficulties that accrue from this location—the difficulty, for example, of forging an unfragmented, unalienated identity, or of discerning resistant positions that can deal effectively with the pitfalls of their emergence from within the folds of dominant colonial ideology—several of these writers (characters, cultural forms), through their elaborations of counterhegemonic themes and strategies, also help in the development of Clare's/Cliff's revolutionary consciousness. And last but not least, the writers, fictional characters, and cultural forms that form a part of Cliff's intertextual apparatus are associated with—indeed, are perceived as representing—particular political positions that are salient to the politics Cliff/Clare wishes to articulate. Invoking them is equivalent to affiliating herself with their politics; invoking them is also simulataneously a way of authorizing her own politics.

These mostly cryptic references, then, not only call upon her readers to identify writers and/or their work and/or a character in this work, but also require them to connect these with the politics the writers, fictional characters, and cultural forms represent. As an example, consider the following reference to C. L. R. James in a postscript to a letter Harry/Harriet writes Clare in *NTTH:*

> P.P.S. Have you ever read C. L. R. James? One impressive bredda dat. I am in the middle of *The Black Jacobins*—I tell you, girl friend, the history they didn't teach us . . . nuh mus' afraid? I think I have fallen in love—with Jean Jacques Dessalines. (146)

Encapsulated in this elliptical reference to James, *The Black Jacobins,* and Jean Jacques Dessalines are several of Cliff's own abiding interests, which include her preoccupation with revisionary accounts of dominant history, enactments of anti-colonial and anti-slavery resistance, and her investment in the revolutionary properties of a Black identity. *The Black Jacobins* is, after all, James's revisionary account of "the only successful slave revolt in history" (James ix). Furthermore, Dessalines's primary loyalty is to the ideology of being Black or of Blackness—not as biologically grounded but as a political construct, a chosen affiliation.

Emperor and first president of Haiti, Jean Jacques Dessalines, unlike Toussaint L'Ouverture, the "hero" of James's *The Black Jacobins,* rejected "things French," recognizing thereby "the temptations of civilization, which for him meant a new, more subtle servitude" (Dayan 74). Simultaneously admired and reviled, albeit by different constituencies, for his positions on land ownership and race, Dessalines, Joan Dayan explains, "in the constitution of 1805 . . . declared that no white, whatever his nation, could set foot on the territory of Haiti as master or owner of property." But

> Certain whites could be naturalized as Haitians: for example, white women who had conceived or would bear Haitian children and those Germans and Poles who deserted Leclerc's army during 1802–3 to fight with the *indigenes.* Further, Haitians, whatever their color, would be known as *blacks,* referred to "only by the generic word *black.*" (77)

Thus Dessalines, Dayan explains, "through linguistic means" sought to resolve ideologically "the most problematic division in the new Haiti [which] was between anciens libres (mostly *gens de couleur,* mulattoes, and their offspring) and *nouveaux* libres (the newly free, who were mostly black)" (77). More importantly, given "what remained of colonial divisions in Saint-Domingue in 1804"—a "tripartite organization" that placed "white masters at the top, servile blacks at the bottom, and freed *gens de couleur* (people of color) in between"—Dessalines, by taking "the lowest rung [made up of blacks] and [making] it a synecdoche for the whole," sought to overturn, dismantle even, "the crucial ideological structure in colonial Saint-Domingue," thereby attempting to eliminate the hierarchy by which "the 'mentality' of servitude [could] be sustained" (Dayan 77–78).

Regarding land ownership, Dayan also remarks on how Dessalines gave "property to those slaves who had, only recently, been considered property." Through his "division of land, [and] attempt to destroy 'false' property titles . . . he tried to carry out what has been called 'an impossible reform of the mentality of the ruling classes, and perhaps his own mentality,'" which, says Dayan, "would make him a favorite of the left-leaning [intellectuals]" for whom, as Anténor Firmin puts it, Dessalines' "actions" embodied "the first manifestation of the sentiment of racial equality" (Dayan 79).

Unpacking the intertextual reference to James, *The Black Jacobins,* and Jean Jacques Dessalines enables the reader to grasp Cliff's politics more clearly. Moreover, in the context in which it appears in *NTTH,* this reference is made by Harry/Harriet to further Clare's politicization. Finally, James (and Dessalines, as Dayan demonstrates) occupies a particular position as a theorist and practitioner of a liberatory, progressive, left-wing politics in general and anti-colonial resistance in particular. Invoking his (their) name, given her own commitment to precisely the politics James and his work (and Dessalines) articulate, both defines and authenticates her project.

Cliff's invocation of Aimé Césaire, his *Cahier d'un Retour au Pays Natal* (*Notebook of a Return to the Native Land*),[24] and through them of the politics of Négritude, in the epigraphs to two section openings of *NTTH* is another attempt at conscious affiliation with a position and a politics designed to explicate and also authorize her own. For one of the most significant political gestures underwriting Clare's coming of age is, as I have been arguing, her coming (in)to an informed and self-conscious understanding of Blackness as a revolutionary identity and consciousness, which is to be distinguished from an unconscious embrace or opportunisitic appropriation of Black identity.[25] The assumptions and ideas underpinning Négritude, particularly the ways in which Aimé Césaire defined them, are crucial in unpacking Clare's/Cliff's understanding and subsequent definition of this identity and consciousness.

A neologism coined by Césaire, Négritude is, according to Janet G. Vaillant, "a theory of black cultural importance and autonomy" (1). Elaborated in Paris in the 1930s by a group of intellectuals from French colonies in close contact with African-American expatriates, Négritude, as most accounts of it point out, was very much a product of the place of its emergence and its historical moment—articulated by diasporic Africans, Afro-Caribbeans, and African-Americans in the imperial metropole, with decolonization yet to come. As Benita Parry observes, Césaire's 1967 interview with René Depestre provides an "explicit reconstruction of Négritude's beginnings":

> We adopted the term négre as a term of defiance. . . . We found a violent affirmation in the word négre and négritude . . . it is a concrete rather than abstract coming to consciousness. . . . We lived in an atmosphere of rejection, and we developed an inferiority complex. . . . I have always thought that the black man was searching for his identity. And if what we want is to establish this identity, then we must have a concrete consciousness of what we are—that is the first fact of our lives: that we are black; that we are black and have a history . . . [that] there have been beautiful and important black civilizations . . . that its values were values that could still make an important contribution to the world. (Qtd. in Parry, "Resistance Theory" 183–184)

Although some exponents of Négritude defined it in essentialist terms, invoking an intrinsic Black identity, for Césaire, as the comments above suggest, it was a willed construction of Blackness as a counterhegemonic identity and consciousness, "a sign," says Parry, "of the colonized condition and its refusal" (182). Thus Césaire is quite right to emphasize that his "concept of négritude is not biological, it's cultural and historical" (qtd. in "The Politics of Neologism" Clifford, 53).

Clare's/Cliff's self-conscious embrace of Blackness is based on precisely such a conception of it as a willed politicized identity and consciousness.[26] In fact, Cliff implicitly and explicitly distinguishes between Clare's self-conscious commitment from Kitty's unself-conscious one and finds the latter ineffectual in the final analysis because it is incapable of being anything other than a private, intimately held passion. (Kitty, we should remember, is viewed by herself and others, including Clare, as possessing an *essential Blackness,* indeed being *essentially* Black, and not only by virtue of her color). In *Abeng,* for example, the narrator notes how "Kitty should have been the daughter of Inez, Mma Alli, and Nanny too." In the absence, however, of her knowledge about them (a knowledge, it's implied, Clare and her creator, Cliff, come to possess), she cannot empower herself or others through it—"using its strength," rather than "protecting" what she construes as her peoples', and, by extension, Blackness's "fragility":

> Kitty's mistake in all of this was casting her people in the position of victim. . . . The revolution had been lost when the first slave ships arrived from the west coast of Africa, and she felt Black people were destined to labor under the oppression of whiteness, longing for a better day but not equipped, Kitty believed, to precipitate the coming of the better day. . . . For all its tenderness, her vision was sad. (128)

Unlike Kitty's but like Césaire's, Clare's/Cliff's understanding of and commitment to a Black identity is self-conscious and mediated additionally by her residence in an(other) imperial metropole: "[B]ecause [Clare] comes to America," observes Cliff in an interview,

> and recognizes racism in the American milieu [becoming obssessed, for example, with the girls killed in the bombing of a Birmingham church during the Civil Rights movement] she can understand better that she comes from a racist milieu. ("The Art of History" 68)

This recognition of racism elsewhere, then, becomes one of the bases for her embracing a "despised" Black identity when she returns to her native land. The awareness that one sees things more clearly and in entirely new ways when one is at a certain distance from it is part and parcel of Négritude's priviliging of a Black identity as well. For example, one of the Nardal sisters, in whose Parisian salon Césaire and Senghor and other intellectuals who were the movers behind the Négritude movement used to meet, connected "the awakening of race consciousness" (also the title of an essay by her) with the migration of these intellectuals—all from the colonial peripheries—to the imperial metropole and to their experience thereafter of "entering a new social environment" that confronted them with "the problem of color as such" (Vaillant 95–97, 90).

5.

Cliff's delineation of Clare's (and her own, for that matter) access to the revolutionary identity Blackness represents as an access acquired through conscious choice and affiliation, and, therefore, as a willed process of politicization has important implications for any discussion of identity as primordial or simply given, and for discussions of resistance/resistant positions that depend on such notions of a pre-given, even biologically determined, identity. At the very least Cliff's delineation calls into question such notions of identity. Of course, given the suspicion with which West Indians—like Edward Braithwaite—regard a "white creole" or "Jamaica white's"—like Michelle Cliff—(in)capacity to represent "the empowering resistance properties of a West Indian text [that] are to be found only in a narrative 'blackness' which operates in direct relation to the writer's subject-position" (Edmondson 181), Cliff's elaboration of (revolutionary) identity as consciously acquired and constructed is both necessary and the means by which she speaks as, and on behalf of, a revolutionary subject(ivity). But in doing so, Cliff neither rejects nor derides nor diminishes organicist, even essentialist, notions of identity. Nor, furthermore, does she suggest that because an identity (and identification) is acquired or constructed, its material location and the specific conditions of its emergence cease to be significant and/or legitimizing. In this regard, then, her work has critical implications as well for those who see identity as only a social construct.

Natal belonging, for example, is an important determinant in Clare's "return" to Jamaica, which represents, as I have noted before, Clare's coming into (full) possession of her revolutionary identity. But this exists alongside and in productive tension with her displacement from Jamaica and her extended stay in the imperial metropole—first the United States and then London, England—which produces, in her case, her conscious understanding of her identity, and its subsequent forms of identifications. In an image paralleling Clare's "return," Cliff, in *NTTH,* describes the "cotta" on which Clare balances her basket as she goes shopping—carrying food she and her fellow guerillas grow on her grandmother's land to barter for food and goods they cannot produce:

> Basket on head, resting on a cotta, bought years ago in a Knightsbridge gallery specializing in African art, carried as a talisman. Now being put to use, its true properties recognized. (12)

The cotta's "true properties" (read: "true identity") are revealed in its "being put to use"—a recognition that itself follows from a (negative) comparison: in the Knightsbridge gallery the cotta served a decorative, not useful, function as an exotic, alien object. "Returned" to its natal space and habitat, it is (and can be) put to (revolutionary) use. The "true properties" of a given object, person, or idea, in other words, are revealed when that object, person, or idea appears in its proper, in this instance its natal, habitat. But this recognition of its natal belonging emerges from a simulataneous recognition of its prior discplacement, its non-belonging, as it were, elsewhere.

Similarly, Clare's "return to [her] native land" as exemplifying her "return" to her proper habitat and chosen identi(ty)fication is premised on her prior displacement—within a habitat and identity she did not recognize or accept as her own. Furthermore, her attempt to transform the desperate conditions in her homeland, and her recognition of the fundamental place of Blackness in this transformation both follow from her displacement, through which, then, her "native land" comes to preoccupy her so centrally. Where displacement yields an apprehension of her "true properties" as a revolutionary subject, her "return" consolidates these "properties," organically binding her to the landscape of Jamaica—when burnt into it, she comes into possession, her narrative implies and Cliff explicitly suggests, of the Blackness she so values. But this image of Clare in the concluding pages of *NTTH* vies with the image of the "actress called in whenever someone was needed to play a Black heroine, any Black heroine," masquerading this time time round as "Nanny, the Coromantee warrior" (206), thereby underscoring the productive tension that is part and parcel of any search for and disclosure of an essentialist identi(ty)fication. The complex ambiguities with which access to Black identi(ty)fication are fraught, as Cliff shuttles between organicist or self-consciously acquired, biologically given or performative conceptions of Blackness, have much to tell us about Cliff's location and the politics she attempts to formulate. They also have much to tell about those situated like her who, through the effort and will to affiliate with positions (and politics) not viewed as legitimately theirs to espouse or represent, can nonetheless aspire to and make their own.

Part III

In *and* of the (Imperial) Metropole

"What a breed of people two hundred years of imperialism has given birth to. If the pioneers from the East India Company could see you. What puzzlement there'd be. Everyone looks at you, I'm sure, and thinks: an Indian boy, how exotic, how interesting, what stories of aunties and elephants we'll hear now from him. And you're from Orpington."

—Shadwell to Karim in Hanif Kureishi's *Buddha of Suburbia*

[T]here was the photograph of my father in my uncle's room [in Pakistan] . . . stories to help me see my place in the world. . . . This was surely part of the way I could understand myself. This knowledge . . . I'd take it back to England where I needed it to protect myself. . . . [But] when I considered staying in Pakistan to regain more of my past . . . I had to think that was impossible. . . . So there was always going to be the necessary return to England. I came home . . . to my country.

—Hanif Kureishi

Chapter 5

"A New Way of Being British": Hanif Kureishi's (Necessary) Defense of Mixtures and Heterogeneity

> *Just as Defoe emphasized the heterogeneous origins of* the Englishman, *we should emphasize his and her heterogeneous existence today, made up of many races and creeds and, within the United Kingdom especially, of nations too. For the positive side of nationalism, that subjective expression of identity and meaning greater than one's own self and family, is in fact complex and particular. In this epoch of migration and miscegenation, it could be that already the majority have multiple loyalties. Perhaps these are necessarily divided ones; mixed, combined, and relative, should a* singular *sense of Britishness be preferable?*
>
> —Anthony Barnett

1.

At first glance, Hanif Kureishi and his work seem a curiously unpromising—perhaps, even singularly inappropriate—choice for exploring an oppositional vision and politics, particularly as these pertain to the hegemony of Western (colonial) ideologies. After all, Karim, the autobiographically informed narrator of his *The Buddha of Suburbia*—a recalcitrant subject at best when it comes to any form of confrontation or militancy—quite candidly and self-deprecatingly admits:

> I was, as a militant, a real shaker and trembler. If people spat at me, I practically thanked them for not making me chew the moss between the paving stones. (53)

Moreover, so often Kureishi's anti-establishtarianism, part of his iconoclasm and irreverance in general, is expressed via a cockiness of attitude and naughtiness in sexual morality, which come across as mere adolescent high jinks, performances of the writer-as-brat. In his interview with Marcia Pally, for example, Kureishi points out how, in *My Beautiful Laundrette,* he wanted to make a film that, eschewing the "dourness and didacticism" associated with this genre (of "ethnic" film, that is), would cause "mass erections" instead (50). More significantly, *The Buddha of Suburbia* (hereafter *Buddha*), his recent *The Black Album* (hereafter *Album*), and some of his nonfictional work, including his screenplays and essays, take on politically charged positions endorsing immigrant ethnic/nationalist identity and identifications within the (imperial) metropole—especially those deemed "right on" or "politically correct"—and instead of attending to their enabling or affirmative function(ing) within specific contexts, subject them to merciless humor and parody, exposing the contradictions or logical fallacies that inhere in them: "We come," as Pally puts it, "to all the Big issues through a few charming but not unnecessarily good people and are thus spared the usual mandates and messages" (50).

What's more, Kureishi draws attention to this aspect of his work by embedding within it criticisms of his critique of these positions by those who are his critique's putative objects. Although his work is dotted with many examples of such self-reflexive critique, I am thinking in particular of Jamila and Tracey's criticisms, in *Buddha,* of Karim's "reactionary" representations of "his own people"—i.e., South Asian immigrants within the metropole. A fellow actor of Karim's in Matthew Pyke's play, Tracey, whom Gayatri Chakravorty Spivak describes as a "politically mature African-British woman" ("The Burden of English" 293), is especially harsh when she takes Karim to task for acting out, as his contribution to the play, his uncle Anwar on hunger strike as Anwar tries to blackmail emotionally his daughter, Jamila, into a traditional arranged marriage with a young man, Changez, from "back home."[1] For Tracey, Karim's attempt to represent Anwar on hunger strike smacks of self-hate because it represents

> "what white people already think of us. That we're funny, with strange habits and weird customs. To the white man we're already people without humanity, and then you go and have Anwar madly waving his stick at the white boys. . . . You show us as unorganized aggressors." (180)

Against Tracey's acutely developed politicized response to the reception among "white people" of Karim's representation of a "minority," politically embattled identity, Karim's bewildered insistence on the irreducible individuality and uniqueness of Anwar ("Indian" he specifies when Tracey categorizes Anwar as "black"; and further "an old Indian man," when Tracey modifies her statement to "black and Asian people" [180])[2] and on the claims of the "truth" (of Karim's portrayal of Anwar) versus (Tracey's attempted) "censorship" does seem to function, for some of us, as a conservative, even reactionary counterargument. However, Stuart Hall's comments on Kureishi's *My Beautiful Laundrette* (applicable to the exchange between Tracey and Karim and to other instances in his work as well) encourages us to look at Kureishi's politically loaded humor and self-reflexive critiques of "right on" attitudes and identifications as an attempt to "open up a continuous critical discourse" about important questions regarding these attitudes/identifications, and not as an attempt to undermine or diminish their saliency. Thus Hall, while commending a

> position . . . which locates itself *inside* a continuous struggle and politics around black [or other forms of minority] representation, but which is then able to open up a continuous critical discourse about forms of representation, the subjects of representation, above all, the regimes of representation,

finds in *My Beautiful Laundrette* one exemplary case of and for the practice of "such a politics of criticism":

> *My Beautiful Laundrette* is one of the most riveting and important films produced by a black writer in recent years and precisely for the reason that made it so controversial: its refusal to represent the black experience in Britain as monolithic, self-contained, sexually stabilized . . . always and only "positive." ("New Ethnicities" 171)[3]

I believe that Kureishi's self-reflexive critiques are attempts also at delineating the inevitable limit(ation)s of any position, progressive or reactionary, that refuses to engage seriously with what for him constitute the complexities and messiness of lived experience and reality.[4] As Shahid, another of Kureishi's autobiographically informed protagonists, finds out, "the world" is "more subtle and inexplicable" (*Album* 143). And, in the context of where he lives and writes, Kureishi finds that:

> if there is to be a serious attempt to understand Britain today, with its mix of races and colours, its hysteria and despair, then writing about it . . . can't apologize or idealize. It can't sentimentalize and it can't represent only one group as having a monopoly on virtue. (Qtd. in Hall, "New Ethnicities" 171)

Concomitantly, if there is to be a serious attempt to understand Kureishi's work, the positions he takes on minority and/or immigrant identities and identifications, and the presence of or ambiguity regarding an oppositional agenda or vision, one must do so by first placing these within the context he invokes and the identification he engages with most substantively: his identity as a "Britisher" and his call for "a new way of being British" ("The Rainbow Sign" 36).[5] In this regard, his work is of a piece with the work of writers and theorists of the Asian and African diasporas whose "originality" proceeds, as Stuart Hall puts it, from "precisely [their attempt to] retell the black experience as an English experience" (quoted in Kobena Mercer, "Diaspora Culture" 51). But in this regard, Kureishi's work is also distinct from the work we have looked at thus far. For despite the fact that James, Rushdie, Aidoo, and Cliff are or have been migrants within (or travelers to) the imperial metropole, their work, particularly the work I examine, though affected by (and itself affecting) their locations as migrants (or travelers) nonetheless focuses on the (de-/neo-)colonized spaces—the homes—they have left behind. For Kureishi, a native-born Englishman whose mother is white and whose father is an immigrant from Pakistan, however, Britain, the imperial metropole, is home, the place to which he inevitably returns ("So there was always going to be the necessary return to England. I came home . . . to my country" ["The Rainbow Sign" 33]). And yet to Kureishi, whose instinctive sense of belonging is with England ("I feel things in England are mine," he says to Pally [53]), England is also deeply inhospitable: In "The Rainbow Sign," he recalls being "racially abused" from a very young age, when he was "five years old" (7). It is in the paradox of his personal predicament, which, to be sure, he shares with other racially marked minorities in Britain—and the manner in which he seeks to resolve it—that one can locate his oppositional vision and politics. They constitute the grounds, in fact, of what many of his readers have recognized as Kureishi's attempt at a reform(ulation) of Britishness.[6] This reform(ulation) insists that, riven as Britishness is with mixtures, which are a product, in large part, as Anthony Barnett points out, "of migration and miscegenation," a "*singular* sense of Britishness" ought to be unacceptable and is in fact inimical to things as they are. Thus, against the myth of a pristine anglo and white British identity, Kureishi offers what Spivak calls the "productive epistemic fracture of the . . . hybrid subject" ("Burden" 293). For this subject (and identity), the rock star Prince serves as one desirable, paradigmatic example:

> He's half black and half white, half man, half woman, half size, feminine but macho. His work contains and extends the history of black American music. (*Album* 34)

In this concluding chapter, I examine Kureishi's reform(ulation) of Britishness—its constitutive elements and, by Kureishi's reckoning, the ethical necessity for undertaking this task. Thereafter, I examine the route through which his autobiographically informed protagonists (and he himself, for that matter) arrive at their self-consciously forged hybrid identities and an awareness of these identities' empowering effects. In this regard, I consider especially the value Kureishi ascribes to role-playing as one of the means by which he and his protagonist, Karim, are able (and instruct others) to construct a hybrid, impure self that Kurieshi finds best positioned existentially and strategically to counter the racism of dominant British society with its demand for a pure (white) subject.[7]

2.

In his introduction to *London Kills Me* (which includes his screenplays for *My Beautiful Laundrette* and *Sammy and Rosie Get Laid,* and an essay on his formation as a hybrid, albeit racially marked, subject, "The Rainbow Sign"), Hanif Kureishi describes fairly precisely the historical moment from within which his critical engagement as a writer, with dominant conceptions of "Britishness" emerges. It is against this backdrop, then, that we should evaluate his counterconception, his dissident vision, as it were, of "Britishness." In the introduction, Kureishi, noting with some surprise the "number of attacks on writers" at the time when *Sammy and Rose Get Laid* was made and significantly when Thatcher and the Tories "were at their most invincible and triumphal," goes on to conclude that these attacks were an integral part of what was really "an argument about language and representation" (x). Margaret Thatcher and the Tories, he says,

> were afraid of anyone who saw Britain as a racially mixed run-down, painfully class-ridden place. For their fantasy was of a powerful, industrially strong country with a central, homogeneous, consensual culture. There necessarily would be hinterlands, marginals, freaks, perverts, beggars, one-parent families, and dissidents. But these didn't matter. There could not be other versions of reality escaping into the world, poisoning the purity of their vision. . . . [I]t wasn't the image of Britain that others should be allowed to see. (x)

Once he perceives the attacks as a function of Thatcher and the Tories' desire to define Britain in a particular and exclusive fashion, he comes to view the "uses of [his kind of] writing and film-making" as a contestation of "the ruling world view": "Writing," he claims, "mattered" (x).[8]

Kureishi's characterization of a Thatcherite (and Tory) conception of Britain as a homogeneous, national space whose integrity is maintained through careful policing and the operation of various exclusionary mechanisms, is underscored by others writing on the same, or similar, subjects.

Drawing on the work of Raphael Samuel and Paul Gilroy, for example, David Morley and Kevin Robins remark on how, in the late 1970s and 1980s, Britain's culture and identity were being reconstructed on "a quite reactionary vision of pastoral England/Albion" from which "blacks [were] systematically excluded" (16).[9] Paul Gilroy's *There Ain't No Black in the Union Jack* demonstrates "how the limits of 'race' [had] come to coincide so precisely with national frontiers." "The effect of this ideological operation," Gilroy adds,

> [was] visible in the the way that the word "immigrant" became synonymous with the word "black" during the 1970s. It is still felt today as black settlers and their British-born children are denied authentic membership on the basis of their "race" and, at the same time, prevented from aligning themselves within the "British race" on the grounds that their natural allegiance lies elsewhere. (46)

In this reconstruction, where "blackness" and "Britishness" were treated as mutually exclusive categories, a common-sense connection was simultaneously assumed, and sometimes explicitly asserted, between "whiteness" and "Britishness."

Underpinning this discrimination between who (and what) genuinely *belongs* within the culture and identity defined as British is the idea that "substantive membership" within the "national community" proceeds from "historic ties of language, custom, and 'race'" (Gilroy 46). For Kureishi, as for others, such a notion of (national) belonging is not only reductive and impoverishing, but also morally bankrupt and therefore ethically untenable. Kureishi finds "the crucial conservative . . . notion" advocating the "company of one's kind"—with "one's kind" glossed as those who share the same "'nation,' . . . colour, race, and background"—a "feeble, bloodless, and narrow" conception of relationships among people that bypasses altogether the "possibilities of love and communication" with those who are not of "one's kind." Embodying the "evil of racism," and therefore morally reprehensible, such a conception is both cause and product of the "denigration of those unlike oneself . . . to the idea of Enemy, . . . Alien, . . . and the Other." More than anything else, it constitutes a denial of humanity, one's own and the other's, because the "failure," whether motivated or unwitting, to connect with others is a "failure to understand or feel what one's own humanity consists in." Thus, according to Kureishi, a "society that is racist" is one that "cannot accept itself, that hates part of itself so deeply that it cannot . . . does not want to see . . . how much people have in common with each other" ("The Rainbow Sign" 28–29).

All this—including the earnest tone and explicit didacticism—is, of course, standard fare in certain strands of liberal discourse on race matters that, in urging an anti-racist politics and mind set, anchor their polemic in a strong belief in the equality and dignity of all human beings. To say this is not to diminish the effect of, or even the impulse behind, Kureishi's argument; it is rather to locate it more precisely within the discourse it draws upon, the ideological assumptions that undergird it, and the audience it directs its appeal to. For here, I think, we would do well to surmise that his audience is composed as much at least of liberal white Britons as any other, for whom the claim that the pathology of racism affects both the victim and the victimizer is an effective ploy because it is part of their own anti-racist positions as well.

Predictably, then, the narrowness, rigidity, intolerance, violence, and lack of humanity that Kureishi associates with "the evil of racism" (of which a reductively framed desire to cleave to "one's kind" is one manifestation), are also characteristics he associates with some of the behaviors and responses resorted to (or devised) to counteract it. Thus, in the same essay, "The Rainbow Sign," in which Kureishi discusses racism directed against blacks, he also describes the "fight back [which] is under way":

> The fierce truculent pride of the Black Panthers is here now, as is the separatism, the violence, the bitterness and pathetic elevation of an imaginary homeland. This is spawned by racism. (27–28)[10]

In *Album,* where Kureishi, with some degree of sympathy and understanding, explores "the path," as Anthony Appiah puts it, "that has led some British Pakistanis to a form of Islamic fundamentalism" ("Identity Crisis" 92), Kureishi's protagonist, Shahid, is nontheless, struck by how "devoid of doubt," "imagination," humor, and an acknowledgment of human frailties these British Pakistanis are in their fundamentalist views (*Album* 120, 161, 143). Shahid begins by admiring their leader, Riaz, but ends by finding him merely "dull and unctous"; "how limited and encased was his mind," Shahid concludes, "how full of spite and acidity" (252). Much like Thatcher and the Tories who cannot tolerate "other versions of reality escaping into the world, poisoning the purity of their vision" (*London Kills Me* ix), Riaz and his fellow fundamentalists cannot entertain alternate accounts of their beliefs and world-view, because they

> would admit no splinter of imagination into their body of belief, for that would poison all, rendering their conviction human, aesthetic, fallible. (*Album* 143)

For Kureishi, if "racism" embodies "unreason and prejudice, ignorance, and a failure of sense," then a response in its own coin also smacks of the same qualities and its advocacy of "separatism, [and] the total loathing of the white man as innately corrupt, is equally unacceptable" ("The Rainbow Sign" 9). Both racism and this kind of response to it represent "insularity, schism, bitterness," denying as they do the possibility of human connectedness and cooperation ("The Rainbow Sign" 37, 28–29).[11]

Quite apart from the substantial ethical problems that cleaving exclusively to either "white" or "black" forms of ethnic/nationalist identifications entail, Kureishi's being and therefore belonging to both prevents him on an existential, even practical level, from treating them as mutually exclusive. Thus, while responding to the "separatism, the total loathing of the white man" by oppressed blacks, Kureishi categorically asserts that he has to "live in England, in the suburbs of London with whites"; that his "mother [is] white," and he's "not ready for separate development," of which he has "had too much . . . already" ("The Rainbow Sign" 9). By the same token, he is not wholly "white," nor can he (nor will he be allowed to) simply assimilate into dominant white British culture. Besides, despite his conviction that "staying in Pakistan . . . was impossible" ("The Rainbow Sign" 33), Pakistan—where "there was warmth and continuity . . . security and much love . . . a sense of duty and community—of people's lives being lived together" ("The Rainbow Sign" 18)—exerts a powerful pull on him. Pakistan, moreover, is a likely source of strength in racist Britain: His "knowledge" of Pakistan, he tells us, "garnered" via "the photograph of [his] father . . . stories to help [him] see [his] place in the world" he can take "back to England" to "protect" himself ("The Rainbow Sign" 33).

As with many expatriate writers, of whom Kureishi's not one insofar as he is "born and bred" in Britain but with whom he nevertheless shares the nostalgic appeal for the "home" country, his view is tinged with a certain romanticism. However, having decided that he belongs in Britain, notwithstanding Pakistan's substantive "place" in his formation, Kureishi unequivocally declares that he is not, he says to Pally, "caught between two cultures"; "I am British," he adds, "I can make it in England" (53). In this regard, he embodies a position embodied by the "second generation and after" of immigrants from the (former) colonies to the (imperial) metropole. Speaking specifically about Caribbean immigrants to Britain, although his conclusions are applicable to immigrants from other colonies as well, Winston James observes:

> Unlike their parents, they compare their position and life-chance within British society . . . with their white British counterparts. Consequently, their experience of racism in Britain is more immediate and their perception of the phenomenon is more uncluttered. . . . Not for them the consolation "It's alright here, basically, *but* we are treated inferior": for them it is "It is not alright here, *because* we are treated as if we are inferior." . . . Unlike their par-

> ents, . . . the second generation of "Caribbeans" [or "South Asians" or "Africans"] are willy-nilly black Britons . . . [who] are . . . by default, creating new cultural forms . . . forging new forms of resistance. (239–40)

For these "black Britons," "blackness" and "Britishness" are profoundly imbricated—a necessary and critical dimension of each other: "Britain and Pakistan" (or, for that matter, Britain and any of the other countries Britain colonized), says Kureishi, "part of each other for years . . . cannot now be wrenched apart" ('The Rainbow Sign" 37). Given this reality, and given, moreover, that this reality follows from Britain's belligerent expansion into various parts of the globe, the onus for achieving this intermixture equitably—the onus, in other words, for bringing about a genuinely multicultural Britain—lies primarily, albeit not exclusively, with dominant white British culture, which must now "learn that being British isn't what it was." However, learning this is not enough for Kureishi; what this knowledge will yield in terms of practice is even more important. Whether, in other words, the identities and identifications along with their ensuing cultural, social, and political forms produced by the "intermixture" that is a necessary offshoot of the former colonizers and colonized inhabiting the same space will be "understood, accepted, humanized" or "violently resisted . . . and characterized by inequality and injustice" is a matter of acute significance in terms of the future of British society. Because unless dominant British society pursues justly the rightful claims of all its peoples, it cannot call itself "democratic" ("The Rainbow Sign" 37).

Of course, in offering this definition of what *should* constitute "Britishness," Kureishi is deploying a familiar weapon in the arsenal of the marginalized or oppressed: i.e., he is enjoining the dominant to live up to its promises—in this case, the promise of its self-definition, its self-mythologizing. For haven't "British society and civilization," argues Kureishi (drawing upon George Orwell's description, in "England, Your England," of quintessential "English" virtues), been markedly characterized by "tolerance" and "gentle manners"? Furthermore, haven't these characteristics been (again invoking Orwell) a "continuous" part of "Englishness," "stretch[ing] into the future and the past . . . something . . . that persists" ("The Rainbow Sign" 35)? Nor is this self-definition of "Britishness" confined to spaces and self-definitions within its national borders. In the formerly colonized territories of its far-flung empire, for example, Britain's "positive legacy" is seen to include "the idea that the health and vigour of a society is bound up with its ability to tolerate . . . a plurality of views [and peoples]" ("The Rainbow Sign" 24).

3.

Through his advocacy of intermixing and intermixtures, which relies on a prior grasp of the powerful ways in which the colonizer and colonized, "Britishness" and "blackness" are intertwined, Kureishi seeks to counteract what Raphael Samuel has characterized as the "born again English nationalism" of the Thatcher years ("Introduction: Exciting to be English" xxxii). Kureishi's reform(ulation) of "Britishness," then, is premised on a culture and condition that embraces, indeed celebrates, impurity, heterogeneity, hybridity. In "Some Time with Stephen," he provides one particularly apt example of an intermixture—a "'hybrid' cultural articulation" that "denot[es]" British "culture's multiple and incongruous accents and cross-cultural inventions" (Parry, "Signs" 13): Music that is a "weird fusion . . . a cocktail of blues and r'n'b shaken with Indian film songs in Hindi . . . Bhangra music! Detroit and Delhi in London!" (179). As black American, reggae, and Indian film music interact and intermesh, enter into and transform each other, in the process creating new cultural forms, presumed national-cultural, and also generic, boundaries are breached, displaced, transgressed in a fashion likely to trouble purists, guardians of authenticity, who view such mixtures and the processes that produce them, as contaminated and contaminating—devolutions from a prior state of (presumed) purity.[12]

Just as hybrid or intermixed cultural forms or articulations, like the one described above by Kureishi, trouble, even threaten, belief in pure or uncontaminated cultural forms, so, too, hybrid identities threaten the belief in a stable, singular identity. Furthermore, as Nikos Papastergiadis puts it:

> hybridity has shadowed every organic theory of identity. Whether it has focused on either the physiological or the cultural constitution of the subject, it has primarily served as a metaphor for the negative consequences of racial encounters. (9)

Tracking the "peculiar historical trajectory of hybridity," especially as it pertains to "a racialized theory of identity" (9), Papastergiadis notes how the negative or positive value ascribed to the hybrid is "always positioned in relation to purity and along the axes of inclusion and exclusion" (10). Thus:

> Whenever the process of identity formation is premised on an exclusive boundary between "us" and "them," the hybrid, which is born out of the transgression of this boundary, will be figured [negatively] as a form of danger, loss and degeneration. (10)

Conversely:

> if the boundary is marked positively—to solicit exchange and inclusion, then the hybrid may well yield strength and vitality. (10)

Another way hybridity and hybrid identity formation gets marked positively or negatively is with reference to conceptions that view identity and identity formation as open and porous or closed, even sometimes hermetically sealed. Hybrid identity and identity formation exemplify the former, but their very openness or porousness is judged as dangerous or threatening by those who subscribe to the latter.[13] Conversely, the latter's closed or sealed character is judged as reductive, narrow and threatening to the openness of the former.

Although Benita Parry is right to caution us about the costs of "hybridisation," especially "hybridisation" that follows from colonialism,[14] in the historical moment and space within which Kureishi is (or finds himself) located, his advocacy of intermixtures and hybridity cannot but be perceived as subversive, if not outright oppositional. For it is against (or in place of) the demand for a pure (white) British subject that he offers the hybrid, multicultural, multiracial subject as the more appropriate bearer of a contemporary "Britishness."

4.

Although hybrid subjects can simply be "born"—offspring, as, for instance, Kureishi and Karim are, of parents of different ethnicities or races—Kureishi's work, especially *Buddha,* suggests that hybrid subjects are more appropriately "made." In the process of being constituted (or constituting themselves) as hybrid subjects, moreover, they acquire (and help others to acquire) a political (and politicized) identity that enables them more securely, and with integrity, to endorse the claims of a genuinely multicultural, multiracial society.

In Kureishi's case (as in Karim's) this making of the hybrid self entails making himself open to or pursuing those parts of himself that connect him with his (thus far repressed or ignored) subordinated or minority identity in Britain—i.e., the Pakistani part of himself, which dominant British culture reviles, and which revulsion he has internalized. In "The Rainbow Sign," for example, Kureishi contradicts Duncan Sandys's assertion that "the breeding of millions of half-caste children would merely produce a generation of misfits" with the assertion that he "wasn't a misfit; [he] could join the elements of himself together" (6). But, in the same essay, he also alludes to moments of ambivalence within himself, of self-hatred, that do not, in fact, exemplify a subject who has been successful at "join[ing] the elements of himself together": "[D]esperately embarrassed and afraid of being identified with these loathed [Pakis]" because "[t]he word Pakistani had been turned into an insult," Kureishi tries actively to "deny [his] Pakistani self." Karim, too, albeit in a more lighthearted and insouciant fashion, while cognizant of the fact that he is "a funny kind of Englishman . . . having emerged from two old histories" (3), begins by identifying more strongly and substantially with the values and claims of dominant British culture and society.

In the opening pages of *Buddha,* for example, Karim discloses his difference from his "Daddio" by observing that after "over twenty years" in Britain, with "fifteen . . . lived in the South London suburbs," his father still "stumbled around the place like an Indian off the boat," whereas "I knew all the streets and every bus route" (7). I see this difference, and Karim's remarking on it, as signifying not only Karim's instinctive sense of belonging within England, but also as a sense that follows from his proximity to, even embeddedness within, dominant culture that, for instance, his father does not seem to possess, even though he has selectively internalized its hegemonic beliefs.

He begins, moreover, by not knowing quite what to do with his (given to him, as it were) hybrid identity, nor does he seem to particularly care. But, like Kureishi (whose alter-ego he is, after all), Karim is, in the course of his travels and travails in *Buddha,* educated (and educates himself) into a recognition of hybrid identity as a valuable tool in defining a politically enabling British identity. His education is cast in a form that is itself an intermixture of two not quite unrelated genres that is worth commenting on. For although my analysis, in its focus on Karim's (self-)education into a fuller, more conscious grasp of a politicized hybrid identity, privileges the *bildungsroman* aspect of the work, *Buddha* achieves its effects by crossing this coming-of-age narrative, modeled on the *bildungsroman,* with the picaresque, or what the reviewer of *Kirkus Review* describes as a "more wayward" and "freewheeling tour through the London of the 1970s—a London as vice- and class-ridden as that of a Hogarth engraving." Both forms contribute to the mapping of a multiracial and multicultural Britain, with the former focused on the (more or less) conscious development of a hybrid psyche, while the latter, with its horizontal spatial arrangement aligned to Karim's insatiable curiosity ("I was the nosiest person I'd ever met" [38–39]), allows Kureishi to sketch in the suburbs and London and their many and varied denizens without privileging any particular class, race, or ethnicity.[15] Thus Gilbert Adair is quite right when he finds in the novel—which he describes as "an utterly unselfconsciously multiracial artefact"—"from surface to psyche, [an] equality [that] is absolute" (34).

The education Kurieshi and Karim (and Shahid in *Album* to some extent) must undergo is weighted towards their acquiring knowledge about, and through that knowledge becoming comfortable with, that part of themselves to which they have had (or allowed themselves to have) little or no access given (their internalization of) dominant versions of a hated and reviled "Paki" identity in particular, but also of minority status in general, Concomitantly, this is an education that dominant British society must also undertake if it is committed to, as Kureishi argues it must be committed to, achieving "a new way of being British."[16] This entails, as Kureishi puts it, a "revitalized and broader self-definition"—which might be "difficult . . . to attain," but must be attempted—that recognizes the substantive claims of excluded or marginalized blacks to genuine national belonging within Britain ("The Rainbow Sign" 37). And it entails a reaffirmation of the much-vaunted but thus-far-observed-in-the-breach British "tolerance" for a plurality of peoples and views.

Well into Karim's coming-of-age narrative—whose "hidden theme," as Margaret Walters so astutely observes, "is Karim's reluctant coming to terms with being Indian as well as British" (16)—Pyke, the director of the play in which Karim plays yet another "Indian," muses:

> "What a strange business this acting is . . . ; you are trying to convince people that you're someone else, that this is not-me. . . . [But] when in character, playing not-me, you have to be yourself. . . . Paradox of paradoxes: to be someone else successfully you must be yourself!" (219–20)

"This," adds Karim, "I learned." But Karim learns something else as well—that the reverse is also the case. For as Karim's (and his father, Haroon's) education in *Buddha* suggests, to achieve "a revitalized and broader self-definition," or to acquire knowledge about a part of the self he has repressed or been denied access to heretofore, Karim must inhabit what he perceives as the "not-me." In his case (and, for that matter, in the case of British self-conceptions), this entails, *Buddha* and Kureishi's other work argues, inhabiting what's been constituted by dominant culture and ideology as a reviled or denigrated minority culture(s) and identit(ies).

Indeed Kureishi's work includes many scenes of instruction wherein imaginatively entering—by watching, reading, role-playing—what is not part of one's self or one's mental baggage provides knowledge considered necessary for growth and survival. Thus, for example, describing the intense impact of the sixties on his ideological formation, Kureishi observes how, by viewing "the sixties on TV," he was "formed by what [he] missed out on" ("The Rainbow Sign" 136). "The Rainbow Sign" also alludes to his "mooch[ing] around in libraries," where he accumulated pictures of, and books by and about, the Black Panthers, James Baldwin, Muhammad Ali, and Malcolm X, all exemplars of insurgent black identity and consciousness, all of whom became part of Kureishi's extended engagement with the dilemmas of minority (black) identification within a society deeply infected by racism.

Although Karim's no reader, "the education system" and he having "split up" (94), and he acquires most of his (self-)education via "his restless trying on of different roles,"[17] Jamila, Karim's dearest friend, almost-cousin, and sometime lover, is, like the Kureishi of "The Rainbow Sign," a voracious reader ("At the age of thirteen, Jamila was reading non-stop, Baudelaire and Colette and Radaguet and all that rude lot" [52]). Through, or because of, her reading, she models herself on Simone de Beauvoir, and role plays at being Angela Davis ("Jammie always carried a photograph of Angela Davis, and she wore black clothes and had a truculent attitude to schoolteachers" [53]), as she "prepared for the guerilla war she knew would be necessary when the whites finally turned on blacks and Asians" (56). Despite his exasperation with her posturing and "right-on" politics and attitudes, Karim cannot help admire (as, indeed, Kureishi expects his readers to admire) how, through her reading and role-playing, Jamila has come into possession of her politically progressive views, her strength of conviction and generosity, all of which are put into practice as she involves herself in anti-racist work and in striving for a better, more egalitarian Britain. Thus, Karim draws attention to her

> feminism, the sense of self and fight it engendered. . . . the things she had made herself know, and all the understanding this gave, seemed to illuminate her . . . as she went forward, an Indian woman to live a useful life in white England. (216)[18]

However, in a work in which the ability to inhabit imaginatively the "not-me" as a way to extend the boundaries of and revitalize the reductively conceived self of dominant culture, it is instructive that both Karim and his father, Haroon, on their way to greater knowledge about themselves and their world, are called upon to inhabit roles that are to begin with dominant British stereotypes about "Indians." For Karim, it is Kipling's Mowgli (with Karim chosen, the director Shadwell cruelly reminds him, "for authenticity and not for experience" [141]). And later, when he is allowed to "choose" his own representation of an "Indian," he sticks to similar dominant conceptions about Indians by playing "an immigrant fresh from a small Indian town," a "wretched and comic character," whose "sexual ambition and humiliation . . . in England" form the substance of Karim's much-appreciated jokes (220). Haroon first plays, then becomes the "Buddha" of the title, by boning up on "Eastern philosophy," "hissing his s's and exaggerating his Indian accent . . . putting back in spade loads" what "he'd spent years" trying to lose (21), just so he can authenticate what is to start with a figure of the Western imagination.

At work here, I believe, is Kureishi's unsentimental and stringent realism about the instruments of one's transformation: the instruments at hand, he seems to be suggesting, are contaminated, shot through with one's desires and needs, and inevitable blindnesses and contradictions; for, after all, the instruments at hand are embedded within the very structures one critiques and hopes to change. Thus, for example, though Haroon locates his antithetical stance to capitalist greed and materialism in his search for his "full potential as a human being," the figure through which he seeks to fulfill this potential, the "Eastern Sage" or the "Buddha," also implicates him in the very processes of capitalistic commodification he deplores. And Karim finds that though "Terry's passion for equality appealed to my purer mind," and "although I hated inequality, it didn't mean I wanted to be treated like everyone else." Like his Dad, who as the Buddha "stand[s] apart," Karim also seeks distinction, and the "power" and the "attention" that accompany it. Hence, while playing Mowgli, "despite the yellow scarf strangling my balls, the brown make-up, and even the accent, I relished being the pivot of the production" (149–50).

Kureishi's clear-eyed recognition of the contradictions and impurities that haunt even our purest, most principled beliefs and gestures is the cause of considerable disquiet and criticism among his detractors—a disquiet and criticism that are reflected, for example, in the reception of his "Asian" or "black" characters and subject matter.[19] However, "in the name of the new politics of reading, as an Asian economic migrant in the United States," Gayatri Spivak, in her reading of Kureishi's *Sammy and Rosie Get Laid,* while cautioning against "construct[ing] an ever-narrowing circle of politically acceptable texts," points toward an important element in Kureishi's approach to politically fraught stances:

> In the article wrongly titled in English as "Commitment," Adorno says of Brecht that his use of montage in *The Resistible Rise of Arturo Ui* simply turns a political problem into a joke. One hopes that Kureishi's montage technique would have satisfied Adorno. It is much more concerned with negotiating a certain kind of unease, a laughter tinged with unease and bafflement. (254)

The knowledge accumulated in the interests of self-development and, following that, of principled political praxis, through the means Kureishi's work proposes, is modest and hedged about with ambivalence, because it is so aware of the elements that undercut it; but it is not negligible, nor should it be cavalierly dismissed. Masquerading as Mowgli on stage, as Karim sends up an "Indian" accent by replacing it with Cockney, he exemplifies what Homi Bhabha has characterized as "mimicry"—not quite a copy of the colonial "original" but its knowing (mis)appropriation, which in turn opens up the space for subversion. Masquerading as an "Indian," moreover, the part-Indian, part-(white) English Karim discovers that the "strange creatures . . . —the Indians— . . . that I'd spent my life denying" were "in some way . . . my people" (212). Through role-playing, Karim has created what he calls "the additional personality bonus of an Indian past"—"important," as Spivak points out, "in the survival technique of the fabrication of a hybrid identity" ("Burden" 292)—without "fetishizing it, as some liberals and Asian radical did" (212). Similarly, masquerading as the "Buddha of Suburbia," Haroon comes to realize that he "remain[s] to all intents and purposes an Indian man," and this realization, if no more than a sign of "the immigrant condition living itself out through [him]" nevertheless, in returning him "internally to India," enables him "at least to resist the English here [i.e., in England]" (64).

5.

In her interview with Meryl Schwartz, Michelle Cliff distinguishes between Clare, "a fragmented character," who "doesn't get a chance to become whole at all" and Harry/Harriet, who is the most *complete* character in *No Telephone to Heaven,*" the "real revolutionary in the book" (Cliff, "An Interview" 600–601; emphasis added). This somewhat more pessimistic or, at the very least, cautious reading of Clare is at odds with the novel's (and Cliff's as well in her "Clare as a Crossroad's Character") affirmation of Clare's "success" in forging an organic connection with her homeland and with (the revolutionary properties of) Blackness. Elsewhere, in an interview with Judith Raiskin, Cliff reiterates her view of Clare's lack of access to her wholeness but also recognizes paradoxically that Clare "chooses" an identity, an "unfragmented" one at that, but this choice is "also a cause of her death"; "it's complicated," Cliff adds ("The Art of History" 64). Further along, in response to Raiskin's question ("Is her choice connected to Harry's choice to become a woman?"), Cliff says: "Yes it is connected to that. Harry tells Clare she has to make a choice. In this world it's a luxury not to choose" (64). Indeed, in *No Telephone to Heaven,* although Harry/Harriet is drawn to the affirmative and emancipatory possibilities of "being neither one nor the other," s/he nonetheless insists on the pragmatic, even political, necessity of making a choice: "For we will be called upon to make our choice. Cast our lot. Cyan live split. Not in this world" (131). Nor is this view a product only of the recognition that identities "in this world" are inevitably delimited by gender, race, class, sexuality and other factors, and that projects devoted to political and other forms of transformation must inevitably work with and through the constraints these impose. This view is also, and more importantly, a product of the kinds of identities/positionalities—"female" and "Black"—Cliff deems resistant.

But Kureishi's novels, screenplays and films, invested in the expansive and emacipatory possibilities of something quite different—a consciously forged hybrid (raced/gendered/sexed) subject as offering the best possible alternative to a normatively defined British identity that is white and emphatically heterosexual—refuse the view that a choice need or indeed should be made. Thus, in *Buddha,* its protagonist, Karim observes:

> it was unusual, I knew, the way I wanted to sleep with boys as well as girls. . . . I felt it would have been heartbreaking to choose one or the other, like having to decide between the Beatles and the Rolling Stones. (55)

Thereafter, he goes on to characterize what would within normatively defined notions of sexuality be conceived of as deviant but that he views as something fortunate, because it provides access to more opportunities and an expanded terrain for his pleasure:

> When I did think about it, I considered myself lucky that I could go to parties and go home with either sex. (55)

Karim's refusal to choose rests on a refusal to confine himself to a singular sexual definition that is itself part of a socially constructed identity. His refusal to choose, in this context, also rests on a refusal to privilege either heterosexuality over homosexuality and vice versa. Finally, here, as in Kureishi's critique of dominant as well as (defensive) minority nationalist identifications, Karim's refusal to choose springs from an equal-opportunity vision that will have no truck with the exclusions that inevitably follow from cleaving to one or another identity and identification. For Karim, as for Kureishi, it's a refusal of an either/or logic in favor of both/and.[20]

What Kureishi and his autobiographically informed protagonists, Karim and Shahid, assert and seek to exemplify is the value of—indeed, the necessity for—a state of radical openness (Shahid, for example, "was determined not to be a closed person" [*Album* 15])—an openness that does not bypass or displace antagonisms, conflicts, and disagreements but that is premised nonetheless on an engagement with different or opposing views that is not dismissive or close-minded. Shahid and Deedee, we are told in *Album,* "argued passionately but without acrimony, both modifying their views" (145).[21]

For those who argue on behalf of a more stringent or limited definition of identity, as, for example, Aidoo, Cliff, and others whose work I have referred to in earlier chapters do, not the least because of its political efficacy and strategic value, the advocacy of openness such as Kureishi asserts is at best premised on utopian expectations, and at worst an evasion, which stems from a witting or unwitting blindness—to the fact of power, to the real inequalities that flow from belonging to one group and not another, one race and not another, one gender and not another, and so on. That these are socially imposed constructions does not denude them of their material force.

For Kureishi, however, the political efficacy of such openness is anchored in its appeal to tolerance—tolerance as a general attitude and mind-set worth cultivating, but also, and more importantly, the tolerance that British dominant culture has always viewed as the cornerstone of its national identity and self-definition. A significant dimension of liberal ideology, the cultivation of tolerance, so consistently appealed to and invoked by Kureishi, presumes as well a constituency of liberal-minded and/or progressive people in Britain, whom Kureishi tries to influence, and whose racism and ethnocentrism—unacknowledged, for the most part, but real nonetheless—he hopes to exorcise. In doing so, Kureishi, like James before him, seeks the inclusion of the marginalized and reviled minority populations within the dominant, but only after the dominant has been transformed, so that the minority populations are not judged by their difference or deviance from the norm, but by their normality and, following that, their right to belong. Thus, speaking about what writing like his can accomplish, Kureishi says:

> You would flatter yourself if you thought you could change things by a film or a play or whatever, but perhaps you can contribute to a climate of ideas. It may be that you write a film about Asian or gay people. . . . [P]eople may see this as being a normal part of everyday life and not a sort of freak show going on in the margins. (Qtd. in Yousaf 20)

Notes

Introduction

1. It's worth noting at the outset, however, that despite significant overlaps among the five writers I examine, Aidoo and Kureishi, for different reasons to be sure, are somewhat of an exception. Unlike James, who, as Caryl Phillips points out, "is at the head of a rich twentieth-century tradition of Caribbean writers who . . . migrated to Britain in order to develop their literary careers" (63), and unlike Rushdie and Cliff, who are also immigrants in Britain and the U.S. respectively, Aidoo has resided in the West only intermittently, albeit for extended periods of time, when she has taught in various metropolitan universities. Kureishi is a native-born Briton; thus, although he is the son of an immigrant Pakistani father and an English mother, he cannot be categorized as an immigrant. Moreover, his entire oeuvre, unlike the texts I examine by the other four, even when it addresses the Pakistani half of his identity does so in the context of what he insists is his British identity. But my inclusion of Aidoo and Kureishi is based on the following rationale: although Aidoo makes her home in Ghana and Zimbabwe interchangably, in the texts I examine she is responding precisely to the pressures generated by her intermittently diasporic position within the metropole. Kureishi's specification of a reform(ulat)ed British identity is premised on the fundamental necessity of incorporating within it the substantive claims that issue from the identities and identifications of immigrants from the Caribbean, Africa, and the Indian subcontinent in Britain.

   Drawing on Fredric Jameson's definition, I use "metropole" to "designate the imperial nation as such" (note 2, 65).
2. Examples of such work, to mention but a handful, include Lila Abu-Lughod's essay on the Awlad 'Ali Bedouin women; Veena Oldenburg's on the courtesans of Lucknow, India; Rolena Adorno's on Guaman Poma; James Scott's on the "weapons of the weak"; and the Subaltern Studies scholars' attempts to record, via the methods of historical retrieval, autonomous peasant consciousness and activity in the context of India. The collections edited by Douglas Haynes and Gyan Prakash, Gary Okihiro, and Dominick LaCapra also contain valuable examples of such work.

3. On the political pressures that impinge on and have helped produce this field, see in particular Gayatri Chakravorty Spivak, especially "Marginality in the Teaching Machine," "The Politics of Translation," "Scattered Speculations on the Question of Cultural Studies"—all in her *Outside in the Teaching Machine*—and "Teaching for the Times." See as well Aijaz Ahmad, especially "Literature Among the Signs of Our Time," "Literary Theory and 'Third World Literature,'" and "Languages of Class, Ideologies of Immigration"—the opening chapters of his *In Theory.*
4. In a characteristically tendentious fashion, Aijaz Ahmad, while examining the conjuncture of events and political pressures that determined the emergence of "Third World Literature" in the Anglo-American academy, remarks on the theoretical manuevers—among them those of "colonial discourse" analysis—through which this literature has been produced as an object of knowledge: "If the whole history of Western textualities, from Homer to Olivia Manning was a history of Orientalist ontology, Third World Literature was *prima facie* the site for liberationist practice" (64). Less tendentiously, Jean Comaroff, in a lecture she gave at the Centre for Modern African Studies, Oxford University, in July 1990, posed the question, Why resistance now? to address the conjuncture that has brought this term to the fore in anthropology, literary criticism, and the social sciences. Thus, she remarked on the moment when conventional social categories of knowledge had begun to come apart, the emergence in academic discourse of examples of the "empire talk[ing] back," and the influence of popular movements as possible "causes." Speaking of analyses of examples of the "empire talk[ing] back," I should add that several of these are prefaced by the claim, such as the one Rolena Adorno makes, that in making these examples visible in academic discourse, the scholar or critic is "decolonizing " knowledge.
5. I have discussed this at some length in an earlier essay, "Reimagining Familiar Dichotomies."
6. In what follows, I offer what I consider the main thrust of their arguments. For Brennan, I restrict myself to his "Cosmopolitans and Celebrities," but by keeping in mind his elaborations of these positions in his *Salman Rushdie and the Third World,* which I will have occasion to address more fully in chapter two. Brennan's more recent *At Home in the World* greatly extends the scope of his discussion of cosmopolitanism by incorporating more writers within its purview and by focusing on cosmopolitanism as a method of analysis that deplores singularity of positioning and binary thinking and invokes instead "intricacy" (86)—complexity/ambiguity—as a definitive value. His preoccupations in this later work, then, are consistent with those of his earlier works. For Benita Parry, I have relied on "Problems in Current Theories of Colonial Discourse," "Signs of Our Times," and "Resistance Theory."
7. Brennan's reading of "Third World" "cosmopolitans and celebrities," based on their favorable reception in the West, is anchored in a suspicion shared

by others. A. Sivanandan, for example, expresses this suspicion in fairly stark and uncompromising terms when, speaking of V. S. Naipaul, he observes: "For the moment 'they' accept you, you are finished, completed; the moment they adopt you, you have sold out, you have become the object of their history, you have no existence apart from them" (33).

8. Like Parry, Abdul JanMohamed also argues against Homi Bhabha's theory of "hybridity" in colonial discourse, because, he says, it "circumvents entirely the dense history of material conflict between Europeans and natives" ("The Economy of Manichean Allegory" 60). He too prefers a view of resistance that proceeds from the assumption that the colonizer and colonized are locked in a Manichean struggle. Both Parry and Jan Mohamed invoke Frantz Fanon in order to explicate their positions. The Frantz Fanon they invoke (and whom, for example, Aidoo brings to mind), however, is quite different from the Fanon Bhabha invokes in support of his own critical procedures (see his "Remembering Fanon") or the one Stuart Hall describes in his "The Afterlife of Frantz Fanon." Fanon, like the forms of resistance he is supposed to embody, is also a product of specific assumptions and procedures of reading.
9. It's important to note that Parry also attends to the strategic and performative dimension in the adoption of nativist positions, especially in her consideration of the Négritude movement in her "Resistance Theory." Or when, in "Signs," she notes that the "construct of binary oppositions, if epistemologically faulty, retains power as a political category" (15). But the emphasis on strategy is neither a consistent nor a primary interest of her analysis. To adapt her own comments on Négritude as a gloss on her procedures, Négritude, she says, is "both an intellectual apprehension of blackness as a construct . . . and a visceral attachment to the powerful fiction of black identity" in which the "visceral attachment" often wins out. As a departure from her usual critical practice, however, consider her essay on Edward Said, "Overlapping Territories, Intertwined Histories."
10. For specific analyses that draw upon Scott, see essays in Haynes and Prakash.
11. For a somewhat different example of resistances enacted by women also anchored in feminist theories and in the work of Scott, see Oldenburg's essay on the courtesans of Lucknow, India.
12. This is Rosalind O'Hanlon's description of the kind of work the Subaltern Studies collective in India has sought to accomplish with respect to analyses of domination. But insofar as this work is undertaken to track and make visible forms of resistance that are entangled with domination, this description applies as well to the forms of resistance.
13. The classic study demonstrating the hegemonic role played by English literature in India is, of course, Gauri Viswananthan's *Masks of Conquest.* See also Susie Tharu and Jasodhara Bagchi, who have analyzed the hegemonic impact of English literature/English studies on "the making of Indian literatures," and "early nationalist consciousness in Bengal" respectively.

14. Part of a characterization of Coleridge by John Stuart Mill, this has been part of my mental furniture at least since my college days in India when I wrote on Coleridge, drawing upon Mill; in this regard, it constitutes one ongoing mediation on my thinking. Describing Coleridge as "'the great questioner of things established,'" Mill says, "with Coleridge . . . the very fact that any doctrine had been believed by thoughtful men . . . was part of the problem to be solved, was one of the phenomena to be accounted for" (1001–1002).
15. With respect to two of the diasporas within the metropole from which I derive my examples, the Caribbean and South Asian, it's important to take into account the following qualifications by Cohen and Clifford: "The first and most evident problem in seeing Caribbean peoples as any kind of diaspora is that they are not native to the area . . . [they] may be thought of as parts of other diasporas—notably the African victim diaspora, the Indian labour diaspora and the European imperial diasporas" (137). Clifford, too, labels it a "quasi diaspora" (306). Nonetheless, Cohen finds that the Caribbean peoples constitute "an exemplary case of a cultural diaspora" because they demonstrate their "cultural retention or affirmations of African identity" and "a literal or symbolic interest in 'return'"; additionally, their "cultural artefacts, products and expressions . . . show shared concerns and cross-influences between Africa, the Caribbean and the destination countries of the Caribbean peoples" (144). On the South Asian diaspora, Clifford observes that this diaspora "as Amitav Ghosh has argued . . . is not so much oriented to roots in a specific place and a desire for return as around an ability to recreate a culture in diverse locations; as such, it "falls outside the strict definition" (306).
16. In recent years, the simultaneous belonging to and estrangement from dominant Western culture that characterizes a diasporic sensibility has been perceived as the basis for a great deal of counterhegemonic work, especially in the Anglo-American academy. Among the public intellectuals who are seen to embody and who see themselves as embodying this position are Edward W. Said, Stuart Hall, and Raymond Williams. By them, see "Media, Margins, and Modernity: A Conversation Between Raymond Williams and Edward Said" and "The Formation of a Diasporic Intellectual: An Interview with Stuart Hall." On them, see Parry ("Overlapping Territories"), JanMohamed ("Worldliness-Without-World"), and Jon Stratton and Ien Ang. On discussions of this as a position that applies to the colonial subject, see Adorno, Bjornson, Pérez Firmat, Irele, and Mouralis.
17. See Linda Alcoff, who suggests that the "concept of woman as positionality" is meant to indicate "how women use their positional perspective as a place from which values are interpreted and constructed," which can produce a "political change in perspective, since the point of departure, the point from which all things are measured, has changed" (434–435). See also note 19, in my chapter two on Rushdie.

Chapter 1

1. The quotations are from McClintock ("Angel" 97), Appiah ("Topologies" 145), and Said ("Yeats and Decolonization" 19) respectively. The phrase "invented traditions" comes from Hobsbawm and Ranger's collection of essays, *The Invention of Tradition*.
2. This is not to deny the existence of another significant strand in discussions of resistance, which, as I point out in the introduction, has in some ways come to represent a privileged position in discussions of resistance. This strand views resistance, its specific strategies and enactments, as affected by and in turn affecting colonial or dominant discourses and ideologies.
3. Khachig Tölölyan, who read and commented extensively on an earlier version of this chapter, which appeared first in *Diaspora: A Journal of Transnational Studies* 2.3 (1993), suggested the title "Inhabiting the Metropole": "'Inhabiting' is important," he said, "because of the links of geography and habitus, because it raises complex questions of the geographical and textual spaces where one dwells"; he also remarked on my "emphasis on C. L. R.'s appropriation and simultaneous arm's length management [of the metropole]" (personal communication).

   In a somewhat similar vein, Belinda Edmondson's recent book *Making Men* (1999) points out how "the tropistic relationship between nineteenth-century Englishness and twentieth-century West Indianness has structured the meaning of 'authorship' and nation in anglophone Caribbean discourse such that what is now recognized as West Indian oppositional discourse to Britain is still marked by a utilization of a specifically English vision of what constitutes intellectual production." She adds: "For the nonwhite, non-English men to make a case for self-government [in the Caribbean], they must state their case as *gentlemen,* which means they must, in essence, be 'made' into Englishmen" (5). C. L. R. James and his work provide her with one of her most extended examples of this tendency within the Caribbean.
4. As a symbol of British "national" life and character that was appropriated by colonized populations, cricket continues to have relevance in contemporary discussions of coloniality and postcoloniality. For example, in *The Tao of Cricket,* Ashis Nandy calls cricket "India's new national game" and analyzes it to disclose its relationship with a number of political, cultural, and social issues: "Some arguments about colonial, neocolonial, anti-colonial and post-colonial consciousness can be made better in the language of international cricket," he says, "than that of political economy."

   > Hence, hidden in the following pages are thumbnail sketches of processes such as diffusion of innovations, transfer of technology, Third World nationalism. . . . Readers may also discover . . . indirect analyses of the subtle psychological processes involved in domination and deculturation—identification with the aggressor, internalization, turning against, reaction-formation, and ritual neutralization of imposed cultural categories. (ix)

Of course, Nandy's position underscores "loss" of "native" categories and values; indeed, he says that the two processes he identifies are linked to his concern about "the problem of dissent in a world where the idiom of dissent is increasingly being defined at the centres of conformity" (ix).

In a bizarre variation that points to cricket's continuing adaptability as a symbol of India's post-independence, "national" life, the Hindu fundamentalist Raman Fielding, in Salman Rushdie's *The Moor's Last Sigh,* conceiving of "cricket as a fundamentally communalist game, essentially Hindu, but with its Hindu-ness constantly under threat," models the "neostalinist inner structures of [his] 'Mumbai's Axis'" on the structure of a cricket team, and presents it as the "change" he must undertake "beyond the boundary" (231).

5. One could say that *Beyond* and *The Black Jacobins* exemplify what Mary Louise Pratt has called the "Arts of the Contact Zone." "Contact zone"—a neologism coined by Pratt—designates the "space of the colonial encounters, the space in which peoples geographically and historically separated come into contact with each other and establish ongoing relationships, usually involving conditions of coercion, radical inequality and intractable conflict." The word "contact" allows Pratt to "foreground the interactive, improvisational dimensions of colonial encounters so easily ignored or suppressed by diffusionist accounts of conquest and domination. A 'contact' perspective emphasizes how subjects are constituted in and by their relations to each other. It treats the relations among colonizers and colonized . . . in terms of copresence, interaction, interlocking understandings and practices, often within radically asymmetrical relations of power" (*Imperial Eyes* 6–7).
6. Jonathan Dollimore's remarks on the "disarticula[tions] and disalign[ments]" through which "the critique of [dominant] ideology" is achieved have some bearing on the "effect" I describe with respect to James's abduction of "Englishness": "In highlighting the contingency of the social, the critique of ideology may also intensify its internal instabilities, doing so by disarticulation or disaligning existing ideological configurations. To borrow a now obsolete seventeenth-century word, the dislocation which critique aims for is not so much an incoherence as a *dis*coherence—an incongruity verging on a meaningful contradiction. In the process of being made to discohere, meanings are returned to circulation, thereby becoming the more vulnerable to appropriation, transformation, and reincorporation in new configurations. Such in part are the processes whereby the social is unmade and remade, disarticulated and rearticulated" (87).
7. Manthia Diawara makes an analogous point; however, his problematic, and therefore his emphases, are different from mine. Drawing upon terms/concepts from film criticism, Diawara focuses on a "self-conscious concept of Blackness as a humanist alternative to Whiteness" (831) with help from which, he says, oppressed Black peoples entered modernity. Addressing the "implications of cricket born outside the logic of Englishness" (838), he

defines "Shannonism" (from the Shannon cricket club) as the mode through which cricket was "wrested . . . away from Englishness" and rendered "pliable enough to serve the needs of the people in Trinidad" (841). I thank Diawara for letting me look at his essay while it was still in manuscript form.

As a wonderful illustration, as it were, of Diawara's claims, Ramchandra Guha points out how James's "masterful study of West Indian cricket, *Beyond a Boundary,* earned him the sobriquet of 'Black Cardus' after Neville Cardus" (a well-known British cricket commentator); thereafter, he proceeds, in the concluding sentences, to reverse the sobriquet: *Beyond* "is a work one would recommend, first of all, to a young cricket lover wishing to build a library of the game's finest literature. But then one must buy a book by Neville Cardus, for Cardus was indeed the 'white James'" (6).

8. In this regard, it is also worth noting Cedric Robinson's comments on James's re-writing in *The Black Jacobins* of Marx and Engels's views about the development of a revolutionary class and consciousness. He underscores how James innovatively redirected their views, broadening their sphere of concerns, certainly, but also insinuating a different explanatory source—in slave culture(s), rooted in the circumstances of their labor and opression—and, thus, trajectory. Also pertinent here are James's comments on why "the independent struggle of the Negro people for democratic rights and equality with the rest of the American nation . . . had to be defended and advocated by the Marxist movement"; "*such independent struggles were a contributory factor to the socialist revolution*" ("Black Power" 233).
9. Specifying intention, as I point out in the introduction, is always a risky and vexing venture. After all "facts" or "evidence" about an author's intentions are not simply "out" or "in" (as the case might be) there waiting to be noticed and retrieved by the insightful critic. Intentions are constructed in order to account for certain positions, certain readings. Having said that, however, I am not willing to give up entirely the notion of *willed* activity on the part of a writer, which can be surmised from close attention to the inclusion of certain materials and not others, certain emphases and not others, and so on, and in this instance from explicit statements as well.
10. Consider, for example, Richard Thomas's remark in his contribution to the special issue of *Urgent Tasks* devoted to James: *The Black Jacobins* "helped us" (i.e., "Black students in a lily-white European survey course") "to challenge this view of white history. We were able to show the historical connections between European economic and political developments and the slave trade, and more importantly for us as Black students, the intimate connection between the French Revolution and the Black Revolt in San Domingo" (121).
11. Consider as an example Foucault's statement: "Where there is power, there is resistance, and yet, or rather consequently, this resistance is never in a position of exteriority in relation to power" (95–96).

12. I am indebted for this formulation to Paul Gilroy, who used it in a talk he gave at the Centre for Modern African Studies, Oxford University, July 1990.
13. Gustavo Pérez Firmat, who has analyzed the impact of a different colonial empire, the Spanish, on the "Cuban condition," also deploys this notion of distance as a way of defining the "critical criollism" (his term for a tendency he characterizes as a "deliberate and nuanced cosmopolitanism") of some of this "condition's" literary products. For him "critical" defines a stance that does not permit "docile emulation of [a] foreign usage . . . it entails [instead] a self-conscious, selective, and sometimes wilful manipulation of literary tradition" (8). In words uncannily like James's, Pérez Firmat finds that the distance that separates the Spanish-American writer from his source "gives him a priviliged hermeneutic position with respect to the cultural products" of Spain (11). Similarly, writing about "Black Art," Sarat Maharaj notes: "It is as if Black Art creates its critique through an initial complicity with the material it denounces, carves out its sense of critical distance by first becoming one with what it intends to keep at arm's length. What we would need to recognize, therefore, is something like an uneven, ceaseless play-off between the oppositional and persuasive modes" (15).
14. For a fuller analysis of Matthew Bondsman's role, see Diawara and also Wynter ("In Quest of Matthew Bondsman").
15. Although in the discussion that follows I focus specifically on how women's presence and/or participation is occluded in *The Black Jacobins* and *Beyond,* when James draws upon some entirely conventional representations to designate their proper sphere of activity, I use gender, which is a relational term, to suggest that any definition/ representation of women is necessarily also a definition/representation of men, that "gender is a way of denoting," as Joan W. Scott tells us, "'cultural constructions'—the entirely social creation of ideas about appropriate roles for women and men . . . a way of referring to the exclusively social origins of the subjective identities of men and women . . . a social category imposed on a sexed body" (1056).
16. With Haiti and the San Domingo Revolution as her example, Joan Dayan, in a subsection of her essay, "Haiti, History, and the Gods," entitled "Défilée," recuperates the presence of "a woman, [who] marched with Dessalines's troops, peddling provisions to the soldiers. Known as Défilée, she was born to slave parents near Cap-François. When she was about eighteen years old, Défilée was violated by a colonist, her master. During the revolution she must have escaped, though nothing is known about her until she became sutler to Dessalines' troops. But with Dessalines' death, Défilée became the embodiment of the Haitian nation—crazed and lost, but then redeemed after desecrating the body of their savior" (86). This subsection, however, is powered by the awareness that even when historians (and narratives of this "only successful slave revolt") mention women,

"we need to consider how these women get mentioned, how their appearances work within the historical narrative. Their stories are something of an interlude in the business of *making history*. Bracketed off from the descriptions of significant loss or triumph, the *blanches* raped and butchered or the *noires* ardent and fearless become symbols for a *la bonte, la ferocité, or la faiblesse*." Thus Dayan questions: "What happens to the unnamed black women during the repeated revolutions in Haiti when mythologized by men, metamorphized out of life into legend? . . . Where did the lived experience of women figure in the demands for black enlightenment? How did they respond to the call for black and mulatto equality?" (90).

17. For a different reading of James's gendered assessments of male and female roles in *Beyond,* see Smith 903–915.
18. My play on James Scott's terms.
19. One of the aims of "post"-colonial scholarship's focus on "alternative" or "revisionary" histories has been to shake or undermine the stability, indeed impermeability, of colonial/Eurocentric history. In James's work this tendency manifests itself, to borrow Said's comments on Fanon, as a "struggle over the same area of experience, culture, history, and tradition hitherto commanded unilaterally by the metropolitan center . . . but now invaded and re-examined critically by a [Caribbean] partisan" ("Third World Intellectuals" 30). The significance of James's work for postcolonial scholarship is addressed in a recent collection of essays, *Rethinking C. L. R. James,* especially in the "Introduction" (1–14) by the editor, Grant Farred, and the essay "The Maple Man: How Cricket Made a Postcolonial Intellectual," also by Farred (165–86).

Chapter 2

1. The critical literature on the politics of location in general is varied and rich. But see especially Linda Alcoff, Ruth Frankenburg and Lata Mani, Chandra Talpade Mohanty, Anuradha Dingwaney Needham, and Kamala Visweswaran (48–50; 100–113; 118–136).
2. In her contribution to the *Satanic Verses* controversy, Feroza Jussawalla dismisses Rushdie as a "totally assimilated and assimilable, Westernized, British-educated Asian intellectual" ("Resurrecting" 114). At an earlier date, in a chapter entitled "Beyond Indianness," Jussawalla observes how *MC* has been read by its Indian readers as a work that exemplifies "the sensibility of Islamic alienation from the rest of India," implying thereby that an "Islamic sensibility" is somehow beyond (i.e., outside) the boundaries of what constitutes "Indianness" (*Family Quarrels* 117). Tim Brennan examines Rushdie as a more fully developed instance of the Third World "cosmopolitan writer." Attached to a given locale in the Third World and thus offering "an *inside view* of formerly submerged peoples for target reading publics in Europe and North America," cosmopolitan writers are those "writers Western reviewers [seem] to be choosing as interpreters

and authentic public voices of the Third World." Despite conceding that these writers offer an *inside view* of the peoples and cultures they write about, Brennan, by privileging the "nation" (and nationalism) as the "empowering image for many Third World intellectuals," finds that writers like Rushdie cannot claim the status of *insiders* fully and authentically because Rushdie, like all cosmopolitan writers, "forward[s] skeptical readings of national liberation struggles," and "thematize[s] the centrality of nation-forming at the same time as demythifying it from a European perch" (*Salman Rushdie* vii–ix; 27–28). KumKum Sangari, who, while remarking on Rushdie's "insider/outsider position as both representative and precarious," notes, "in *Midnight's Children,* Rushdie attempts to play with . . . two ways of seeing" that draw upon "indigenous resources" and an "epistemological mode" based on western modernism. "However," she goes on, "somewhere in this region where two ways of seeing intersect lies the curious desire of the narrative to confirm both ways of seeing." This leads Sangari to question whether Rushdie's project "can . . . be said to be itself transformative, to effect a different mode of understanding, to remake the reader" (236–39).

3. For a sampling of Rushdie's assertions about writers in general and his own work in particular, see his essays "Outside the Whale" and "The Location of Brazil," and his interviews with Rani Dharker, David Brooks, and with *Third World Book Review,* entitled "A Dangerous Art Form." When Rushdie talks about a writer's adversarial relationship with dominant ideologies, he certainly does not mean all writers, only writers like himself. Thus his remarks serve as a gloss on what he conceives of as the nature of his work.
4. These remarks were made in response to the following statement and question from *Third World Book Review:* "You live in England and write in English, but your two major novels have been centered on the Indian subcontinent. In what way has this influenced your work?" His response about being an "insider and outsider in society" is interesting precisely because he leaves "society" unspecified—which one does he have in mind? The Indian, which he's left behind but which he writes about? The British, within which he now lives? Both?
5. Abdul JanMohamed, in an analysis of "border intellectuals" (with whom Rushdie's migrants have much in common), compares the "syncretic [border] intellectual" with "his specular counterpart" to suggest how the former is "more 'at home' in both cultures" than the latter, who "while perhaps equally familiar with two cultures, finds himself or herself unable or unwilling to be 'at home' in these societies and . . . subjects the cultures to analytic scrutiny rather than [like the syncretic border intellectual] combining them." As an exemplary instance of the specular border intellectual, JanMohamed presents Edward W. Said, "who, [as] neither quite an exile nor quite an immigrant, is able to develop, out of his more complicated border status, an enabling theory of 'exile' as an 'ascetic code of *willed* homelessness'" ("Worldliness" 97, 99). JanMohamed sees Rushdie as an example

of the syncretic border intellectual. I argue, however, that Rushdie occupies a position similar to the one JanMohamed ascribes to the specular border intellectual, and want therefore to suggest that there might be more overlapping between the two categories than JanMohamed is willing to concede. By distinguishing so sharply between the two, JanMohamed does not seem to take into account an important dimension of migrant experience. Meaghan Morris's comments on Paul Willemen (also an immigrant to Britain) are appropriate here, as they note how the "border is a dense and busy place in Willemen's writing": "Emigration is a frictional experience; to *inhabit* another language as well as a different culture is not only to encounter barriers, resistances and gaps, but to live and work everyday in chafing proximity *with* them." "A cultural politics," then, "needs something that a radical rhetoric does not; namely a realistic way of negotiating differences between *overlapping* social spaces" (7–9).

6. Teresa Ebert's comments on "rewriting" are apropos here:

> Rewriting is a (post)modern strategy for . . ."activating" the "other" suppressed and concealed by dominant modes of knowing: it articulates the unsaid, the suppressed, not only of the texts and signifying practices, but also of the theories and frames of intelligibilities shaping them. Voicing this silenced "other" displaces the dominant logic—dislodging its hegemony and demystifying its "naturalness"—and unleashes an alternative potential. (888)

7. I use the term "deconstruct" deliberately to evoke some of the meanings Gayatri Chakravorty Spivak ascribes to "deconstruction." Consider especially her remarks on Derrida's study of Nietzsche, *The Ear of the Other* ("Feminism and Deconstruction" 129). Elsewhere Spivak notes:

> "deconstruction is not most valuably exposure of error, certainly not other people's error. The most serious critique in deconstruction is the critique of things that are extremely useful, things without which we cannot live on, take chances" ("In a Word" 4).

8. For this discussion of the formal features and ideological effects of re-play, I am indebted to Lawrence Needham, who first discerned it in the passage from Rushdie I have quoted, and elaborated upon it in subsequent conversations while I was writing this chapter.
9. Reflecting perhaps the symbiotic relationship that obtains between Nehru's and Rushdie's narratives of nation, Indira Gandhi, in her 1980 "Foreword" to a reprint of her father's *The Discovery of India,* observes how Nehru's *Autobiography* "has been acclaimed as not merely the quest of one individual for freedom, but as an insight into the making of the mind of new India." Rushdie's *MC,* also published in 1980, literalizes this claim through its central trope: Saleem as India, India as Saleem.
10. Brennan is right to underscore Rushdie's skepticism about and demystification of the projects of nationalism and nation-formation as these pertain

to the Indian subcontinent. But in situating it as a product of Rushdie's cosmopolitanism, which entails his "artistic embrace of the West," as it were, Brennan's analysis disregards the imbrication of nationalism itself and of narratives of nation, such as Nehru's, for instance, with Western, specifically colonial British histories and their representations of India. For this one need only glance through Nehru's footnotes to realize how much (and many) of those narratives and representations are incorporated, often uncritically, into his *Discovery.* On the imbrication of nationalist narratives and their representations of the subcontinent with those of their colonial counterparts, see Partha Chatterjee's (especially *Nationalist Thought*) and Gyan Prakash ("Writing Post-Orientalist Histories").

11. Although while discussing specific narrative strategies within *MC,* I point to Saleem as the source of these strategies, it is not easy or even feasible to maintain too rigid a distinction between his narrator and Rushdie. In the marketing of *MC* and in its reviews, as I observe in "Author(iz)ing *Midnight's Children* and *Shame,*" readers have assumed that *MC* is autobiographical—that Saleem's life coresponds not just with that of his nation, but also with that of his creator. Rushdie's own comments, indicating that he is not averse to readers assuming this correspondence, point out how "Saleem and Salman are, after all, if you look back etymologically, kind of versions of the same name. . . . [Saleem's] the same age as me more or less . . . he grows up in my house, he goes to my school" (*Midnight's Children* and *Shame*" 12–13).
12. Most of the time in this chapter, by nationalist I mean anti-colonial nationalism.
13. The etymological link between Aadam and Adam is surely a deliberate ploy on Rushdie's part to suggest a narrative of "origins"—the "origins," in this instance, of Saleem in a narrative whose central trope links Saleem's birth and life with that of his nation's. But Aadam is not Saleem's real grandfather, since his daughter, Amina Sinai, is not Saleem's real mother. Amina's real son, Shiva, is replaced by Saleem by Mary, who later becomes Saleem's ayah. Saleem is actually Methwold's son by Vanita. By implication, then, the genealogy of this nation is based an "abduction" and a "mixture"—of the "indigenous" with the "foreign."Yet, given Rushdie's focus on India's neocolonial present, this genealogy is not altogether inappropriate.
14. Gyan Prakash, who finds nationalist historiography "complicitous" with Orientalist procedures (a complicity, one would argue, signaled in *MC* via Saleem's use of the same *topos* about India's primeval identity by both colonialist and nationalist representations of the subcontinent), notes that "burdened with the task of articulating an anti-colonial, national view," nationalist historiography "could not but be different from Orientalism"; at the very least, it had to transform India from "passive to active, inert to sovereign" ("Writing Post-Orientalist Histories" 388).
15. Tariq Ali, for example, notes that Scott's *Quartet* "must be considered the main forbear of *Midnight's Children*" (88). Similarly, Brennan locates echoes from both Forster and Scott in Rushdie's work (82).

16. I borrow this from Spivak's definition of deconstructive practice.
17. Saleem's re-staging of Methwold's staging of his—and by implication the British Raj's—departure from the Indian subcontinent is anticipated almost a hundred pages earlier, through a representation whose power to disrupt British self-descriptions of their departure, however, is less a product of interrogation and more of challenge and contestation through a symbolically loaded statement. It aspires to the status of an overtly oppositional representation, not the least because at its center is an idiosyncratic rather than "borrowed" cultural detail—*paan*-chewing men involved in an intricate game of Hit-the-Spittoon—and Indian protagonists. Nevertheless, the method of its presentation is continuous with Saleem's other "shots" of India. Framed on the one end by the figure of a "whistling [optimistic] doctor," Aadam Aziz in 1942, the year of the Quit India movement, this "shot's" other end describes the town's military commander passing by in his army staff car as the old men hit the spittoon with "long red jet[s] of betel juice." As the car "knocks over the spittoon" the "dark red fluid . . . like a red hand in the dust of the street . . . points accusingly at the retreating power of the Raj" (45).
18. In his interview with David Brooks, Rushdie makes this comparison explicit when he notes: "One of the things I've discovered is that migration has a kind of suppressed history, somewhat like the way women have been erased from history" (67). Ranajit Guha, recommending that we "develop the special skills and . . . disposition" that will enable us to hear the "small voices" of history otherwise repressed by dominant ("statist") History, privileges "women's voice" as a particularly significant "[small] voice of defiant subalternity": "[W]omen's voice," he says, "once it is heard, will activate and make audible the other small voices as well" (3, 12).
19. The distinction I have in mind draws upon Sandra Harding's comment, which itself draws upon her analysis of standpoint theory: "If I must learn to understand my life from the standpoint of, say, third world women's lives, it is not so that I can speak *as* or *for* third world women. Men must learn to see themselves as they appear from the perspective of women's lives, but it is not that men can speak *as* or *for* women." The consequences of doing so, that is, "starting thought—theorizing—from women's lives," according to Harding, "decreases the partiality and distortion in our images of nature and social relations" (123–124). Consider as well Linda Alcoff's definition of "women as positionality": "the concept of women as positionality [itself a product of social construction, not an essentialist definition of what women are] shows how women use their positional perspective as a place from which values are interpreted and constructed" ("Cultural Feminism" 434). The crucial differences that result from this positionality are not necessarily a product of "any new facts about the world," although new facts about the world might indeed be disclosed by virtue of occupying "women as positionality." Rather, the differences are a product of a "political change in perspective since the point of departure, the point from which all things are measured, has changed" (434–435).

20. For a more fully developed account of Rushdie's problematic representation of women, see Inderpal Grewal and Sara Suleri.
21. I focus on this essay beause it includes and adds to what Rushdie has said or written in other essays and inteviews. Relevant page numbers for these are as follows: "The Indian Writer in England" (76–79) and "Outside the Whale" (137–138); interviews with Rani Dharker (356, 360) and David Brooks (59, 67).
22. On this, see especially Ania Loomba and Suvir Kaul, who caution us against viewing "migrant experience" in monolithic terms, and urge instead our attention to its internally differentiated nature because it is marked, for example, by class, race, and gender. Rushdie's *Satanic Verses,* in fact, does show the migrant experience and sensibility as heterogeneous and internally differentiated; historically dynamic, their lines of unity with and division against a normatively construed center or margin are constantly open to negotiation and reworking. Loomba and Kaul also caution against a too easy conflation of the "migrant experience" with "the entire experience of postcoloniality" (13–14).
23. See, for example, "In a Word" 8–9.

Chapter 3

1. The epigraph to Frantz Fanon's *Black Skin, White Masks* (hereafter *BSWM),* these are Aimé Césaire's words from *Discours su le Colonialisme;* Fanon's text was originally published in 1952 in French as *Peau Noire, Masques Blancs.*
2. In *BSWM,* Fanon addresses the situation affecting colonized Antilleans from Martinique before decolonization. Aidoo's works are located in Ghana after it was decolonized. Ghana gained its formal independence from Britain in 1957. During the course of an interview, Aidoo told me: "I started working on [*Our Sister Killjoy*] in '72." It was completed while she was on a "trip here to the United States when [she] came over as a Consultant Professor . . . [in] '74-'75. By the time [she] was leaving for Ghana, the book had taken shape" (131). Her short stories, entitled *No Sweetness Here,* collected over a decade, were published in 1970.
3. Like Fanon, Amilcar Cabral, leader of the liberation movement against Portuguese colonialism in Guinea-Bissau, asserted how "in order not to betray [the] objective [of national liberation], the petty bourgeoisie has only one choice . . . [it] must be capable of committing suicide as a class in order to be reborn as revolutionary workers" ("The Weapon of Theory" 110).
4. By "analyzing" the "massive psychoexistential complex" that has been caused by "the fact of the juxtaposition of white and black races," Fanon "hope[s] . . . to destroy it" (*BSWM* 12). Similarly, Aidoo views her "responsibility" and "commitment" as a writer as entailing the task "to unfold or open for ourselves and our communities what exists, what's wrong, the problems" ("An Interview" 130).

5. Given Fanon's relentless invocation of white and black as binary opposites, it's difficult to view his vision as other than Manichean. Nevertheless, it behooves us to attend to the following caution from Stuart Hall: "Since the text [*BSWM*] so remorselessly returns to the binary oppositions, *black/white, coloniser/colonised,* I wonder how many of his readers unconsciously slip into reading him as if binaries are the exclusive focus of his tale? As if the real title of his book was 'Black Skin, White Skin'? Ignoring the fact that though his subject is, of course, framed throughout by the dichotomous and Manichean structure of racism as a binary system of representation and power, it is the split of the divided self, the two sides within the *same* figure—the colonial Negro—which centrally preoccupies him" ("The Afterlife of Frantz Fanon" 18).
6. Christopher Miller observes that "the character of African nationalism before independence is peculiar—'nationalism' in this period refers not to affiliation with a particular *state* but simply to any form of resistance to colonialism and any explicit assertion of African rights" (*Nationalists* 120). Despite the fact that Aidoo's narratives are set for the most part in post-independence Ghana, when she refers to "African nationalism," she is "using the term to embrace the global African World" for which, as she puts it, "Pan Africanism" is the "better term" ("An Interview" 124)
7. On anti-colonial, nationalist historiography's use of inversions, see Gyan Prakash, who analyzes the case of Indian nationalist historiography and demonstrates its inversion of Orientalist histories of India.
8. Vincent Odamtten identifies a "broad spectrum of Ghana's elite (the petit, national, and comprador bourgeoisie) and the literate working and peasant classes" (note 14, 176) as Aidoo's "primary audience," whose education and abilities at self-criticism Aidoo's work hopes to inculcate (12).
9. On this reversal, see also Ranu Samantrai and C. L. Innes.
10. It is hard to determine whether or not Sissie, whose journey from Africa to Germany and back constitutes the central narrative strand in *OSK,* is also the narrator, or even whether there is only one narrator. Nevertheless, although Aidoo, in her interview with me, cautioned against viewing Sissie as an autobiographical subject, or even as someone whose values and judgments are identical with Aidoo's, insofar as Sissie's views and judgments are not subjected to either implicit or explicit interrogation, we must view them as those *OSK,* and its creator, Aidoo, endorses. Moreover, though the subtitle to the work, *Reflections from a Black-eyed Squint,* presumably specifies the perspective from which its evaluations and judgments are formulated and forwarded, that perspective itself is not held up for analysis, interrogation, or criticism.
11. On this issue, I disagree with Ranu Samantrai's claim that *OSK* "provides an example of how a non-racialist, non-foundational African identity might lead to Pan-African solidarity" (140). What's more, I don't think this claim quite coheres with her analysis that supports Aidoo's advocacy of a "nationalism which is based on an African essence" (141).

12. The phrase "'Look, a Negro,'" reiterated several times in this chapter relentlessly drives home the point that the "negro" is discursively constructed "by . . . the white man, who had woven [the negro] out of a thousand details, anecdotes, stories" (111). Fanon also adumberates this construction's powerfully debilitating material effects—imprisoning, objectifying, producing "Shame and self-contempt. Nausea" (116). However, where Fanon exposes how the "negro" (and his concomitant "blackness") is constructed as an object through his analysis of "the history that others have compiled for me" (120), Aidoo in deploying a reverse discourse, whereby whiteness is rendered negative through its analogy with "pickled pig parts," seizes upon an instrument from the arsenal of the oppressor to achieve her effects.
13. Mary Louise Pratt, whose *Imperial Eyes* offers a definitive analysis of European travel writing and its intimate connections with European expansionism, notes how "arrival scenes . . . serve as particularly potent sites for framing relations of contact and setting the terms of its representation" (78).
14. In his introduction to George Jackson's prison letters, Jean Genet's remarks on using the oppressor's/enemy's language transgressively seem apropos with respect to Aidoo's use of the analogy linking "white(ness)" to "pickled pig parts": "the prisoner must use," says Genet, "the very language, the words, the syntax of his enemy, whereas he craves a separate language belonging only to his people." (Aidoo's Sissie also "craves" a "private" language—one, that is, not burdened with or exemplifying her people's "enslave[ment]" so that "the messengers of [her] mind always come shackled" [112].) "It is perhaps a new source of anguish for the black man," Genet continues, "to realize that if he writes a masterpiece, it is his enemy's language . . . which is enriched. . . . He has then only one recourse: to accept this language but to corrupt it so skilfully that the white men are caught in his trap. To accept it in all it richness . . . to suffuse it with all his obsessions and all his hatred of the white man. That is the task" (qtd. in Dollimore 99–100).
15. However, one could agree with the general truth of June Jordan's remarks, made in a different context, and argue that the harshness and cruelty of this analogy reflects less a retaliation than a *response,* and an appropriate one at that, to a prior (even if unreflected upon) cruelty. In *Civil Wars,* for example, Jordan notes:

    > The courtesies of order, or ruly forms pursued from the heart of rage and terror or grief defame the truth of every human crisis. And that, indeed, is the plan: To defuse and deform the motivating truth of critical human response to pain. . . . If you make and keep my life horrible, then when I can tell the truth, it will be a horrible truth; it will not sound good, or look good, or God willing, feel good to you either. There is nothing good about the evils of life forced into useless and impotent drift and privation. There is

very little attractive or soothing about being strangled to death, whether it is the literal death of the body or the actual death of the soul that lying, that the humiliation, and the evil of self-denial guarantees. Extremity demands, and justifies an extreme response. Violation invites, teaches violence. Less than that, less than the absolute cessation of normal events in the lock of normal duress is a lie, and worse than that, it is a blasphemous ridicule of the self.

16. In a humorous reworking of Fanon's notation of the objectified nature of "blackness," Galahad in Sam Selvon's *The Lonely Londoner's* ruminates on how the color "black" is at the source of his troubles: "'Lord, what it is we people do in this world that we have to suffer so?' . . . And Galahad watch the colour of his hand, and talk to it, saying, 'Colour, is you that causing all this, you know. Why the hell you can't be blue, or red or green, if you can't be white? You know is you that cause a lot of misery in the world. Is not me, you know, is you!' . . . So Galahad talking to the colour Black, as if is a person, telling it that is not *he* who causing botheration in the place, but Black. . . . Galahad get so interested in this theory about Black that he went to tell Moses. 'Is not we that people don't like . . . is the colour Black' . . . 'that is a sharp theory; why you [Moses] don't write about it'" (88–89). This rumination, it's worth pointing out, is provoked by an encounter identical to the one Fanon's chapter, "The Fact of Blackness" proceeds from: "'Mummy, look at that black man!' A Little child, holding to the mother hand, look up at Sir Galahad" (87).
17. Mary Louise Pratt's analysis of Mungo Park's *Travels,* in focusing on scenes of "mutual appropriation" (81) and "reciprocal vision" (81–85), qualifies somewhat this view of all European travel writing as appropriative and overdetermined by colonial agendas. While conceding that "European ideologies are never questioned directly" (84), she nevertheless finds that "[i]n comparison with a great many other travelers, especially some of the Victorians who followed him, Park affirms plausible worlds of African agency and experience. His relational approach to culture raises genuine possibilities of critical self-questioning" (84).
18. Ranu Samantrai also links what she calls Aidoo's "necessary nationalism" with "the urgent need to oppose neocolonialism" (155); Vincent Odamtten has subtitled his book on Aidoo (*The Art of Ama Ata Aidoo*) as *Polylectics and Reading Against NeoColonialism* as a way of signaling Aidoo's salient interests. With respect to Aidoo's emphasis on building a new Africa, consider her dedication in the opening pages of *OSK* to "Nanabanyin Tandoh,/who knows how to build;/ people/ structures/ lives . . ." Here "build" resonates with the attempt to create something new, worthwhile, and transformative.
19. Under the rubric of "culinary critique," Ranu Samantrai demonstrates how Aidoo connects "the abundance of food available [in Germany], huge portions of which are served to the visiting students" with the "'sweet foreign aid' fed intravenously, secretly to otherwise comatose African rulers."

"These students," Samantrai continues, "represent the next generation of the 'Diplomats/Visiting Professors/Local Experts in sensitive areas' of their countries. Their overfeeding [therefore] is both drugging and bribe, an attempt to create another dependent generation convinced of the inevitability of Europe's self-appointed role as wealthy benefactor of the world" (151). Samantrai's gloss on *OSK* alerts us to an important theoretical influence on Aidoo's work, that of dependency theory, which emerged, says Frederick Cooper, "from Africans' and Africanists' disillusionment with the fruits of independence in the 1970s" (1518). Neil Lazarus, who has studied the "configuration . . . the set of relations . . . between radicalism and increasingly totalizing theory at the end of the 1960s and between such theory and political paralysis" in the work of a fellow Ghanaian writer and contemporary of Aidoo's, Ayi Kwei Armah, notes how "deriving from the analysis of Latin American societies, but, by the mid-1960s, being applied also to the ex-colonies in Africa and Asia, dependency theory set itself the task of accounting for the intensifying structural dependence—above all, and seemingly paradoxically, within the era of postcolonialism—of the 'peripheral' Third World upon the 'center' constituted by the capitalist First World." Rejecting as "ethnocentric and unhistorical" First World invocations of "traditionalism" as the cause of the failures of independence, "dependency theorists maintained that the condition of the 'Third World' should be viewed not as one of *undevelopment* but rather as one of *underdevelopment* . . . the forcible 'peripheralization' of the 'peripheries'" (*Resistance* 196–197).

20. Noting how "African people have been [and continue to be] under all kinds of onslaught," and that "African people are at the bottom of the human pile," Aidoo insists that "whatever is left for us to recoup cannot be done unless we see ourselves as a people, as a nation"; and again: "We have to rescue ourselves as *our own people* . . . we need *us*" ("An Interview" 125, 126).
21. These statements by Aidoo appear as blurbs on the back cover of the Longman African Classics edition of *OSK*.
22. During the past several years many rich discussions of various nationalisms as fundamentally gendered discourses and practices have become available. See especially the essays in the following collections: *Nationalisms and Sexualities, Recasting Women,* and *Scattered Hegemonies;* see also Partha Chatterjee's *The Nation and its Fragments,* Susie Tharu and K. Lalitha's introduction to *Women Writing,* and Kumari Jayawardena; and see essays by Susan Geiger, Anne McClintock, Gayatri Chakravorty Spivak, and E. Frances White. On nationalism in general, the list of works is formidable; I have found the following collections helpful: *Nation and Narration, Becoming National,* and *Nationalism, Colonialism, and Literature.* See also Partha Chatterjee's *Nationalist Thought* and essays by Parry and Lazarus.
23. As a case in point, consider a story like "Two sisters," also in *No Sweetness Here,* in which the protagonist, Mercy, a female member of the native

bourgeoisie, against the vociferous objections of her older sister, Connie, has an affair with a member of Parliament, Mensah Arthur, so that she can acquire all sort of luxuries otherwise not available to any but the ruling elite. When, following a coup, he lands up in jail, Mercy promptly dumps him and transfers her attentions to a captain in the new regime so that despite changes in political power, she may continue to enjoy the standard of living she's become used to. Nor does her sister escape unscathed. Her acute disapproval of Mercy's "selling" of herself is suspended as soon as she needs something from Mensah Arthur, hitherto unavailable sewing machine parts that Mensah Arthur can procure for her using his influence as a member of Parliament.

24. While discussing African writers writing in the first decade of independence, among whom he includes Aidoo, Neil Lazarus notes how they were "immediately recognizable" because they were "all concerned in their work, to the point of obsession, with independence as failure, with what independence did *not* bring, with the unraveling of the social unity that they firmly believed had been in evidence at various ceremonies of independence" (*Resistance* 22).

Chapter 4

1. Not literally, in that Cliff already possesses speech and so doesn't need to journey towards it, but metaphorically, insofar as she perceives speechlessness or the lack of speech as a product of her alienation from an identity she views as her own but that she has been taught to despise (*Claiming an Identity I Was Taught to Despise* is the title of one of her collections): "I could speak fluently," says Cliff, "but I could not reveal." To accomplish the latter, she says, she needed to "approach [herself] as subject," and that entailed getting past the "message of anglocentrism, of white supremacy," of years of "indoctrination," so that she could come into a history and an identity she could recognize as her own (*The Land of Look Behind* 12–13).
2. Cliff uses Black interchangably with African and dark (or darkness). I have selected Black from this range because it evokes a politicized identity and identification, and, in Catherine MacKinnon's words, "a heritage, an experience, a culture and personal identity" (116)—all of which I believe Cliff has in mind. I follow Cliff's lead in using upper case for each time I use Black. I retain the lower case where it appears as such in quotations from others.
3. Belinda Edmondson glosses "white" further in her note: "I use 'white' here in its social and political sense; unlike the United States, where racial categories are rigidly drawn, in the West Indian context they are more nebulous; as such people of black and white heritage can be called 'brown,' 'red,' or 'white' depending on the particular circumstances. Though from all evidence Cliff regards herself as black or mixed, in Jamaica she is considered, socially and economically speaking, 'white'" (181).

4. To my mind, there is no such thing as an unmediated identity (or identification). Even Aidoo's work, which assumes an organic, and thus unmediated, connection between nation and identity, nevertheless also treats identity and identification, as I suggest in the preceding chapter, as strategic and therefore mediated, and not simply biologically grounded. Notions of authenticity and inauthenticity premised on biologically grounded notions of identity, however, continue to haunt discussions about national, gender, racial, sexual, and other identities and forms of identifications as, too, they specifically haunt the reception (and production, for that matter) of Cliff's work.
5. Françoise Lionnet also points out how Cliff "is involved in an 'archaeological' enterprise not unlike that of Michel Foucault in his *Archaeology of Knowledge,*" but she restricts it to Cliff's attempt to present the "subject formation" of Creole culture.
6. This subject continues to preoccupy Cliff, for in her recent work, *Free Enterprise,* she devotes an entire section not to histories of colonial successes but to accounts of resistance against them. Thus in one of the stories told by a Hawai'ian man the murder of James Cook by the natives is recalled and we are told, "his murder, and he was definitely murdered, was retribution for [his giving the natives syphilis], as much as for other matters" (49); another, by a Tahitian, recounts Fletcher Christian's dastardly acts and how "one warm Pacific night . . . the slave men rose as the moon rose, and sliced Mr. Christian's throat across, and some other throats, and set themselves free" (57).
7. Stella Dadzie, writing on slavery and resistance in Jamaica, confirms this insight into women's erasure from accounts of resistance: "From the earliest European descriptions of Maroon intransigence to latter day accounts of slave rebellions," says Dadzie, "black women are conspicuous by their near invisibility or by their apparent collective propensity for promiscuity and collaboration with the planter class." Dadzie views this erasure as typical of the limitations of (patriarchal) definitions of women's roles and of a view of resistance that can only conceive of men as heroic, and that perceives "organized male violence as its single or most valid expression" (21).
8. Addressing the fit between form and content in this and other recuperations of Jamaica's forgotten histories within Clare's coming-of-age narrative, Lionnet observes how these "reconstruction[s]" are "translated through what Glissant has called 'économie parcelaire,'" whereby Cliff

   > has recourse to a textual economy of "small plots" that seems to correspond to the economy of "small plot farming" that maroon slaves used to engage in. Because she wants to claim the heritage of the maroons who survived in large numbers in Jamaica, Cliff uses a narrative fragmentation that is but the mimesis of another form of cultural and economic dispersion and segmentation. In order to survive in the high mountain regions of the island, the maroons would cultivate small plots of land that were alternatively

cleared out and left fallow because they were always on the move and to avoid being captured. (335)

9. Representing part of Clare's patrilineal ancestry, Judge Savage, her great-great-grandfather, "concerned about the survival of his race" and afraid that through miscegenation the "white seed would be diluted and the race impoverished" (38) preferred, on the eve of emancipation, to "burn the hundred slaves on his plantation" (71).
10. Dadzie also comments on the "coercive agro-industrial labour" in Jamaica as "undoubtedly one of the most extreme and potentially dehumanizing systems ever devised" (22).
11. For example, in an interview with Opal Palmer Adisa, Cliff observes: "I've always been struck by the misrepresentation of history and have tried to correct received versions of history, especially the history of resistance" (280).
12. Lionnet's characterization of Cliff's reference to a history "sunk under the sea, or scattered as potash in the canefields" (339).
13. I borrow this formulation from Amilcar Cabral's "The Weapon of Theory"; the full quote reads as follows:

    the essential characteristic of imperialist domination [is] the negation of the historical process of the dominated people. [In response] national liberation is the phenomenon in which a given socio-economic whole rejects the negation of its historical process . . . national liberation is the regaining of the historical personality of that people, its return to history through the destruction of the imperialist domination to which it was subjected. (102)

14. This comment, by Cheikh Anta Diop, quoted by Bogumil Jewsiewicki, himself quoting Paulin Hountondji quoting Diop, points to the pervasive belief among intellectuals from the colonized parts of the world of the fundamental importance of a people writing their own history—writing as subjects of their own history.
15. Here it is worth noting that the Blackness Cliff valorizes, and whose histories she therefore seeks to recuperate, when viewed—as it is by some Afro-Caribbean critics of Cliff—as biologically grounded is ironically deployed to characterize Cliff and her work as "inauthentic" or "compromised" by virtue of her "compromised location" as a "Jamaica white" or "white creole" (see Edmondson 180–182).
16. Along similar lines, Edmondson, remarking on how "a significant percentage of today's West Indian writers are women, and [that] they have recast the familiar project of articulating a national identity to reflect their specific experiences as women," finds Cliff's "novels critical to working out the problems of race and gender in this regard." Cliff's novels, she says "reflect her search for an Afrocentric identity through her matrilineal ancestry . . . thus, the search for a black history/identity is intimately bound up

with a latent feminism as well as with a revolutionary social consciousness" ("Race" 182).

17. *Abeng* and *NTTH* are often read as narratives of nation "parallel[ing]," as Helen Maria Lima puts it about *NTTH,* "the formation of the young self to that of the developing nation" (35). I read these works somewhat differently; at the same time as they tell us about what the fledgling nation, Jamaica, has become or is, the trajectory of Clare's development, which they are assumed to parallel, points towards a desired definition, not yet achieved, and thus toward what the nation should become, rather that what it already is.
18. In *Abeng,* Cliff remarks: "Color is of course often metaphorical," a point that is reaffirmed and dramatized with considerable force in the following conversation from *NTTH:*

    > One night, the man who serves the table, who attends meetings on his evenings off, unable to hold his tongue any longer, turns to the people at the table, and says, "You know, Mas' Williams, when we get de power, de power fe de people, t'ings not gwan be easy fe de white smaddy of Jamaica dem. Per'haps wunna and wunna family should t'ink 'bout de hemigration. Wit' all respect, sah." "I never did hear such rudeness, Faith." The master speaks across the server to the mistress. "We are not white, Joshua, so we are not worried." The mistress taking pity on the houseboy and deciding that his words came from benign concern. Flattered nonetheless that even this hignorant countrybwai did t'ink she white. Not understanding his use of metaphor. (20)

19. Thus, when Kitty returns to Jamaica, she leaves Clare behind in the U.S. with her father while she takes Clare's younger sister, Jenny, "back home" (*NTTH* 84).
20. Addressing the debilitating consequences of the "complex hierarchy of human shades"—a veritable "multilayered pigmentocracy"—that flourished (indeed, still flourishes) in the Caribbean, whereby "the coloured demeaned and humiliated by whites . . . had few qualms about pouring scorn upon . . . the African section of the population," which, in turn, internalized this scorn as "self-doubt and self-contempt," Winston James, in his analysis of "the making of black identities" in Britain, notes how "British racism has helped to create an identity—which perhaps under different circumstances would not have developed—among Afro-Caribbeans living in Britain. The crude bipolar (black/white) allocation of the ethnic groups within Britain has helped severely to undermine the color hierarchy which vitiated the Caribbean psyche" (233; 230).

    Along somewhat similar lines, Manthia Diawara defines blackness as a phenomenon that emerged in the West, not in Africa, as a mode of opposition against the domination of peoples of African descent in the context of slavery and colonialism (Diawara 831). Cliff's staging of the choice that faces Clare (and herself) as a choice between Black and white resonates, I think, with some of the same concerns.

21. While I do not think Clare and Cliff are identical with each other and should not therefore be treated as interchangeable with one another, from here on a great deal of what I analyze does, in fact, apply to both of them, thereby making them virtually interchangeable for the purposes of my analysis. Hence this awkward designation, which I prefer nonetheless to reducing one to the other, or subsuming one under the other.
22. For my formulation here I am drawing upon and adapting Terry Eagleton's comments in "Nationalism: Commitment and Irony"; the relevant passages are as follows: "to undo the alienation [produced by social class], you ha[ve] to go not around class, but somehow all the way through it and come out on the other side" (22). And: "For the past is of course what we are made of; and the impasse of any transformative politics is that it can unravel what Marx and Stephen Dedalus call the nightmare of history only with the poor, contaminated instruments which history has handed it" (27).
23. Cliff's designation, in her essay "Clare Savage as Crossroads Character," signifying the opposing, conflicting elements that exist within Clare:

    > Her name . . . is intended to represent her as a crossroads character, with her feet (and head) in (at least) two worlds. Her first name means, signifies, light-skinned . . . and light-skinnedness in . . . the island of Jamaica in the period of British hegemony . . . stands for privilege, civilization, erasure, forgetting. . . . Her surname is self-explanatory. It is meant to evoke the wildness that has been bleached from her skin, understanding that my use of the word *wildness* is ironic. (265)

    In what is perhaps not an unintended irony, *Abeng* makes clear that her first name is, in fact from the black girl, Clary, even though Kitty allows Boy and Clare to think otherwise, and it is her father, surnamed Savage, who, because of the whiteness he so assiduously cultivates, tries to bleach Clare of all her wildness.
24. The project (and methods) of Césaire's *Notebook of a Return to the Native Land* also haunts another of Cliff's works, "Obsolete Geography," a section of *Claiming an Identity They Taught Me to Despise* included in *The Land of Look Behind*. Analyzing the "problematic of translating space in Aimé Césaire's work," Indira Karamcheti remarks on "his struggle to name and inhabit the local place—Martinique rather than a French colony, the Caribbean rather than *outre-mer*—and to make them Black." This struggle, she adds, "conducted in the dominant codes of French," makes of Césaire's *Notebook* both a "deconstructive and a constructive project": "He deconstructs the cult of the exotic which transforms the Caribbean—the place and its inhabitants—into commodities fetishized by the Europeans" and "simultaneously . . . construct[s] a Caribbean subject by creating a myth of origins, a genesis, and a genealogy." In so doing, Césaire's *Notebook* "enacts

a linguistic reoccupation of the land . . . building an incremental mass sedimented out of the repetition of the specifics of place" (181–83). "Obsolete Geography," alluding to, and detailing, the geography, flora, fauna, and inhabitants who people the Jamaican landscape—"nam[ing]," that is, and "inhabit[ing] the local place"—participates in a similar project, its "cartographic impulse" at once creating anew by deconstructing colonial Jamaica.

25. My analysis thus far (and in what follows) attempts to distinguish (and distance) what I consider Cliff's informed and self-conscious understanding of and participation in Blackness as a revolutionary identity and consciousness from the charges of opportunism leveled against her. Edmondson succinctly summarizes these charges when, referring to the debate surrounding Cliff's explicit attempts "to revalue black identity in her novels," she notes how many "West Indian feminists and intellectuals . . . feel that even as Cliff is described as a black feminist novelist in America (where she lives and writes) her novels are not truly part of an Afrocentric Caribbean discourse because her project as a feminist emanates from an American feminist sensibility and perhaps more importantly that her discovery of a black identity is a foreign fashion that she has appropriated" ("Race" 181–82).

26. Cliff's delineation of Clare's (and her own) willed politicization—her delineation, in other words, of the process through which this politicization is achieved—is comparable to Stuart Hall's recollection of "the long, important, political education of discovering that I am 'black.'" "Constituting oneself as 'black,'" he adds,

> is another recognition of self through difference: certain clear polarities and extremities against which one tries to define oneself. We constantly underestimate the importance, to certain crucial political things that have happened in the world, of this ability of people to constitute themselves, psychically, in the black identity. . . . The fact is 'black' has never been just there either. It has always been an unstable identity, psychically, culturally, and politically. It, too, is a narrative, a story, a history. something constructed, told, spoken, not simply found. . . . Of course Jamaica is a black society, [people] say. In reality, it is a society of black and brown people who lived for three or four hundred years without ever being able to speak of themselves as 'black.' *Black is an identity which had to be learned and could only be learned in a certain moment.* In Jamaica, that moment is the 1970s" ("Minimal Selves" 116).

Chapter 5

1. In "Locating the Asian Experience," Perminder Dhillon-Kashyap characterizes the obsessive focus of the British media, in particular, but others as

well, on arranged marriages in South Asian communities as the deployment of a stereotype through which the British media and others seek to disclose the inherent "primitive[ness] of South Asian culture" (122).

2. Although referencing Kureishi's strong investment in individualism, which at the same time marks his investment in liberalism, Karim's insistence here also expresses a position in a debate on the internal difference that "Asian identity" represents within a larger, more embracing concept of "black" identity and identification. Addressing the "uniqueness" of the "culturally different beliefs and practices" of "South Asian immigrants to Britain," Tariq Modood, for example, argues against an anti-racism anchored in conceptions of "black" identity alone: "[I]n Britain," he says, "there was also a second cleavage: a West Indian-Asian dualism. Despite the different political and cultural histories that this cleavage represented, British anti-racism, having accepted the first opposition between black and white, continued to deny any political or anti-racist strategic significance to this internal division." "This denial," for Modood, constitutes a "politically naive falsification which has had a deletrious effect on the ability of Asians to mobilize for anti-racist struggles" (156–58).
3. In a similar vein, Kobena Mercer, noting the "unexpected scale of [*My Beautiful Laundrette*'s] popularity," also remarks on the audiences who have "actively disliked the film": "the Asian communities," for example, whose "angry reactions focused on the less than favorable depiction of some of the Asian characters which, when read as emblematic of the community, were seen as reinforcing certain ethnic stereotypes"; or Professor Norman Stone, who disliked it "for portraying a negative image of contemporary England." For Mercer, as for Hall, this "contradictory reception" is a function not only of different readings of the same cultural text, or of "conflicting ideological viewpoints" produced by an "ethnically diverse society," but also of how "image-making has become an important arena of cultural contestation" ("Recoding" 20). A word about the designation "black" in Hall's comment: "Black" during "the seventies and eighties" in Britain was a "political rather than a racial category" (Mercer, "Recoding" 22), "coined," according to Hall, to point to the "common experience of racism and marginalization in Britain," and in turn "to provide the organizing category" for resistance among people otherwise differentiated by their histories, ethnic identities and heritage ("New Ethnicities" 163–64). Kureishi, as note 2 above makes clear, both invokes such a designation and refuses, or at least complicates it, as, for example, in the argument between Tracey and Karim in *Buddha*.
4. Speaking against perceptions that reduce an individual to a sexual or racial type and/or an ideological position(ing), Kureishi, in his interview with Pally, in characteristically outrageous language, avers: "Men say women are stupid or that their sexual organs will devour them. Women argue that all men are rapists or the source of sexism is the penis. Whites say blacks are stupid or worthless; blacks counter that white men are devils. So no one

has to look at why people become the way they are or notice that people resist their training and change" (50). It's instructive that Kureishi links the ability to change with the ability to resist socialization into "naturally" given positions or types.

5. Kureishi uses "Britisher" and "Englishman" (and Britain and England) interchangeably—the latter more frequently than the former—to describe himself and the place where he lives. But Raphael Samuel's attempt to discriminate between the two is worth rehearsing here, if for no other reason than to indicate that some of the meanings that cluster around the former are arguably more productive for the project Kureishi undertakes. Samuel notes:

    > "English" . . . [has] all kind of pleasant connotations. It evoke[s] a people rather than a state. . . . Because of its association with the language, it [is] umbilically tied to English literature. Because of its subliminal association with the countryside. . . ."English" is smaller and gentler than "British" [which is] . . . hard rather than soft and belonging to specific historical epochs rather than the timeless one of "tradition." It is a political identity which derives its legitimacy from the expansion of the nation-state. Its associations are diplomatic and military rather than literary, imperial rather than—or as well as—domestic. . . . [I]t allows for a more pluralistic understanding of the nation, one which sees itself as a citizenry rather than a folk. It does not presuppose a common culture and it is therefore more hospitable to newcomers and outsiders. ("Preface" xii)

    On this see also Linda Colley, Bernard Crick, and Hugh Kearney.
6. On this, see especially Simon Gikandi, Radhika Mohanram, Gayatri Chakravorty Spivak, and Nahem Yousaf.
7. Although referencing a different culture and historical moment, Nikos Papastergiadis, in an essay that attempts a genealogy of the hybrid, mentions Gilberto Freyre's specification of the "hybrid's advantage" with respect to "establish[ing] a firm grounding for 'racial democracy'" that is worth recalling here: "In this new celebratory myth," says Papastergiadis, "which was defined in opposition to the polarities of race relations in the U.S., hybrids were conceived as lubricants in the clashes of culture, they were negotiators that would secure a future free of xenophobia" (10).
8. Characteristically, Kureishi ironizes, or at the very least casts a self-deprecating glance at, even this source of his oppositionality. In "Some Time with Stephen: A Diary," which is also included in *London Kills Me,* Kureishi remarks on his collaborator Stephen Frears's exasperation that, though in *My Beautiful Laundrette* he and Kureishi were making a film critical of Thatcher's England, most likely Thatcher would approve of their entrepreneurial skills through which they were able to make a box office success of a very low budget project (130–31).

9. Gilane Tawadros is right to remind us that "blackness" and "Britishness" have been "fixed and mutually exclusive categories" at least "since the mid 1950s when settlers from West Indies and the Indian subcontinent arrived in Britain in significant numbers" (41). Hanif Kureishi's work, however, while referring to this history of discrimination, is more specifically focused on Thatcher's Britain with its resurgent imperialism (on which see as well Salman Rushdie's "The New Empire in Britain" and "Outside the Whale"). In a similar vein, Yasmin Alibhai observes: "In her Bruges speech, Mrs. Thatcher's recent invocation of a common experience rooted in colonial history of the European nation-states" was premised "confidently and unashamedly" on "'the story of how Europeans explored and colonised and—yes, without apology—civilised much of the world,'" which story Thatcher went on to characterize as "'an extraordinary tale of talent, skill, and courage'" (quoted in Morley and Robins 16).
10. *Album* dramatizes the causal relationship between the two through an account of Chad, who is one of the more militant British Pakistanis to embrace Islamic fundamentalism, by Deedee Osgood, who is Shahid's mentor and lover: "Chad," she tells Shahid,

    > was adopted by a white couple. The mother was racist, talked about Pakis all the time and how they had to fit in. . . . Chad could hear church bells. He'd see English country cottages and ordinary English people who were secure, who effortlessly belonged. You know the whole Orwellian idea of England. . . . [T]he sense of exclusion practically drove him mad. He wanted to bomb them. (116–17).

    Cognizant as Kureishi is of the fact that "insisting on [one's cultural] identity" represents a plea for belonging and the desire not to "always . . . be on the outside of everything" (*Album* 15, 170), he is also insistent that because they stem from a response that is primarily defensive, inwardly directed, and exclusionary, they become the very thing they abhor and set out to counter in the first place.
11. However, in *Buddha,* Karim specifies how difficult it might be to control responding in kind in such a racially fraught context:

    > And we pursued English roses, as we pursued England; by possessing these prizes, this kindness and beauty we stared defiantly into the eyes of Hairy Back [Karim's nickname for the racist father of one of his girl friends], into the eye of [his] Great Fucking Dane. We became part of England and yet stood proudly outside it. But to be truly free we had to free ourselves of all bitterness and resentment, too. How was this possible when bitterness and resentment were generated afresh everyday? (227)

12. On this see also Jean Fisher.

13. However, for Jean Fisher, drawing on comments from Sarat Maharaj, "hybridity" is not entirely "open," nor does it entirely escape reductionism:

    Sarat Maharaj points out [Fisher notes] . . . that in its popularity, [hybridity] risks becoming an essentialist opposite to the now denigrated "cultural purity." Hybridity is, moreover, a term still fraught with connotations of origin and redemption: two identifiable entities combine to produce a third capable of resolving the contradictions of its origins. (6)

    However, the conditions (and identities) Kureishi describes, for which I have used the terms intermixture and hybridity, resemble more closely Parry's definition of "'hybrid' cultural articulation[s]" as "denoting culture's multiple and incongruous accents . . . descriptive of subject positions and social conditions traversed by heterogeneous cultural inflections" ("Signs" 13). On hybridity, also see Werbner and Bauman.
14. Drawing on Manthia Diawara's critical interrogation of hybridity, Benita Parry notes that

    in glossing Edward Brathwaite's definition of creolization "as one's adaptation to a new environment through the loss of parts of oneself and the gain of parts of the Other," Manthia Diawara—who cautions that the question of hybridity "as *the* correct way of being Black in the West has enabling elements as well as uncanny moments"—has observed: "one must be aware of the fact that in fusing Whiteness with the seductiveness of hybridisation, one is sacrificing not only a part of Blackness, but certain Black people." ("Signs" 20)

    Given Kureishi's location, however, this caution applies selectively.
15. Kenneth Kaleta notes that the title for the "novel's first draft dated 27 December 1987" was *The Streets of My Heart,* which Kureishi later changed to *The Streets of the Heart;* "the[se] two earlier titles," Kaleta adds, "reflect Kureishi's focus . . . mak[ing] paths of the human heart and the grid map of the city synonymous, an emotional A to Z" (63).
16. In a related fashion, Spivak, while commenting on the role of Channel 4 of British Independent Television, for which Kureishi and Frears's film, *Sammy and Rosie Get Laid,* was produced, notes:

    Channel 4 . . . is interested in something like what Gramsci calls the formation of the intellectual, the preparation of viewership. Thus the role of this television channel is also didactic. If part of the role is to educate the so-called minorities, part is to educate the so-called dominant viewership about the minorities as well. Stephen Frears's films fit this latter role more persuasively, I think. (244)

17. Margaret Walter's review of *Buddha,* which attends to what Alamgir Hashmi, in his essay on the novel, has called "both a structural and sym-

bolic device" (26), observes: "Karim's job as an actor neatly underlines his restless trying on of different roles through friendship and sex—with Charlie, with the political hardliner Terry, the upper-class Bohemian Eleanor and the trendy director Pyke" (16). To her list one might add the roles he inhabits via his stint in the theater—Kipling's Mowgli, and Changez, who is Jamila's husband and Karim's almost-brother-in-law.

18. The characterization of Jamila could well be an anomaly in Kureishi's normal procedures. For though Karim initially makes various satirical digs at her "right on" politics, she is also, over the course of the novel, protected from the disclosure of contradictions and hypocrisies that the other characters are heir to. In this, her character comes to function as a type, an ideological positioning, her "feminism," coming to define what is strong and moral and essentially good about her. In her characterization, in other words, Kureishi seems to settle for something more programmatic than he allows himself with respect to the other characters.
19. For example, consider Perminder Dhillon-Kashyap's concern that in *My Beautiful Laundrette* Kureishi has "created a new victim, the white fascist—a victim of economic circumstances who is being exploited by petty bourgeois Asian businessmen" (125); or Gurinder Chaddha's complaint that Kureishi has used "the Asian part of himself . . . without real cultural integrity" (quoted in Yousaf 17). bell hooks, although appreciative of Kureishi's "determination to be open and frank about his take on reality, which means that he shows us not just those neat little politically correct areas of his vision but also those murky, shadowy, confused places," is nonetheless troubled by and quite harsh about Kureishi's representation of race matters in *Sammy and Rosie Get Laid,* as, too, she is critical of the "stylish nihilism" of Kureishi's film, whereby "difference—rather than becoming the new site for resistance and revolution, for ending domination—becomes the setting for high spectacle, the alternative playground" (163).
20. Radhika Mohanram's commentary on *Sammy and Rosie Get Laid,* which makes a similar point with respect to Kureishi/Frears's subversion of normative notions of sexuality, is worth quoting from at some length:

    > In a scene, [as] Sammy and Rosie get ready for a party, [they] are mimed by [the lesbian couple] Rani and Vivia. The intimacy of the lesbian couple is juxtaposed with that of the heterosexual. Interestingly, there is a reversal of point of view. That is to say, normally it is the dominant group, or the heterosexual couple who are positioned as voyeuristic . . . [which carries with it] associations of othering and the objectification of that other. But here it is the lesbian couple who are voyeuristic. . . . Rani's amusing interpretation of heterosex "where a woman tries to come but can't, and a man tries to stop himself but can't" in fact others heterosex and positions homosex as not only the norm but also one of fulness and tremendous satisfaction. . . . Rani's dismissal of heterosexualtiy gains meaning, however, only within that polarity.

> That is, for homosexuality to achieve its fulness of meaning and semblance of identity, it must include heterosexuality. . . . The scene continues and Rani's comment that Rosie calls this household "hedgehog" because "there are so many pricks around" is met with Sammy's . . ."not all pricks are men" . . . which destabilizes any strict adherence to polarities within representation. If women can be pricks just like men, homosex . . . [is] not a privileged category. . . . [W]ithin Kureishi's films sexuality split into polar opposites is seen as incomplete. (128–29)

21. Harriet Gilbert's review of *Buddha* captures well the radical openness I suggest Kureishi and his protagonists value and try to exemplify in their behavior. Finding the "looseness" of the plot in *Buddha* appropriate because "Kureishi isn't interested in tying up neat little packages," Gilbert observes:

> On the contrary, in pursuit of meanings of racial identity and racism, of culture clash and post-imperial adjustment—of the ways in which these things may be honestly, usefully, be written about—he's willing to open up every parcel in sight. Left-wing political idealism, Muslim cultural traditionalism, the intricate, hidden colonialism of liberal, white intellectuals: these and more get unwrapped and examined with beady, respectful eyes. And it's this combination of real, impatient curiosity with what I can only describe as critical tolerance that gives the book force, not only as a novel, but as a political sign-post: an indication of how we might break through our smugness, misunderstanding, and anger. (2)

Bibliography

Abel, Elizabeth. "Black Writing, White Reading: Race and Politics of Interpretation." *Critical Inquiry* 19 (Spring 1993): 470–498.

Abu-Lughod, Lila. "The Romance of Resistance: Tracing Transformations of Power Through Bedouin Women." *American Ethnologist* 17 (1990): 41–55.

Adair, Gilbert. "The Skin Game." *New Stateman and Society* 3 (March 30, 1990): 34.

Adorno, Rolena. *Guaman Poma: Writing and Resistance in Colonial Peru*. Austin: University of Texas Press, 1986.

Ahmad, Aijaz. *In Theory: Nations, Classes, Literatures.* London and New York: Verso, 1992.

Aidoo, Ama Ata. *Our Sister Killjoy; or Reflections from a Black-eyed Squint*. Essex, U.K.: Longman Group Inc., 1977.

———. *No Sweetness Here.* Essex, U.K.: Longman Group, Inc., 1970.

———. "An Interview with Ama Ata Aidoo." By Anuradha Dingwaney Needham, *Massachusetts Review* 36.1 (Spring 1995): 123–133.

Alcoff, Linda. "The Problem of Speaking for Others." *Cultural Critique* (Winter 1991–1992): 5–32.

———. "Cultural Feminism versus Poststucturalism: The Identity Crisis in Feminist Theory." *Signs: A Journal of Women in Culture and Society* 13 (1988): 405–436.

Ali, Tariq. "Midnight's Children." *New Left Review* 136 (1982): 87–95.

Appiah, K. Anthony. "Identity Crisis: Is the Hero of this Novel English, Pakistani, Muslim, or All Three?" *New York Times Book Review* v. 100 (Sept. 17, 1995): 92.

———. "Out of Africa: Topologies of Nativism." LaCapra: 134–163.

Bagchi, Jasodhara. "Shakespeare in Loin Cloths: English Literature and the Early Nationalist Consciousness in Bengal." Joshi: 146–159.

Baker, Houston A. Jr., Manthia Diawara, and Ruth H. Lindeborg, eds. *British Black Cultural Studies: A Reader.* Chicago: University of Chicago Press, 1997.

Barker, Francis, Peter Hulme, and Margaret Iverson, eds. *Colonial Discourse/Postcolonial Theory.* Manchester and New York: University of Manchester Press, 1994.

Barnett, Anthony. "After Nationalism." Samuel, vol. 1: 140–155.

———. "Salman Rushdie: A Review Article." *Race & Class* (Winter 1985): 91–98.

Bauman, Zygmunt. "The Making and Unmaking of Strangers." Werbner and Modood: 46–57.

Bello, Cindy Rose. "Fanon, Aidoo and the Forging of a Pan-African Nationalist Solidarity." Unpublished essay.

Bhabha, Homi K. *The Location of Culture.* London and New York: Routledge, 1994.
———. *Nation and Narration.* London and New York: Routledge, 1990.
———. "DissemiNation: time, narrative, and the margins of the modern nation." *Nation and Narration:* 291–322.
———. "The Commitment to Theory." *Questions of Third Cinema.* Ed. Jim Pines and Paul Willemen. London: British Film Institute, 1989: 111–132.
———. "Remembering Fanon." Foreword to Frantz Fanon, *Black Skin, White Masks.* London: Pluto Press, 1986.
———. "Signs Taken for Wonders: Questions of Ambivalence and Authority Under a Tree Outside Delhi, May 1817." *Critical Inquiry* 12 (1985): 111–132.
———. "Of Mimicry and Man: The Ambivalence of Colonial Discourse." *October* 28 (1984): 125–133.
Bjornson, Richard. "Alienation and Disalienation: Themes of Yesterday, Promises of Tomorrow." Mudimbe: 147–156.
Bracey, John. "Nello." Buhle: 125.
Brennan, Timothy. *At Home in the World: Cosmopolitanism Now.* Cambridge, MA: Harvard University Press, 1997.
———. *Salman Rushdie and the Third World: Myths of the Nation.* New York: St. Martin's Press, 1989.
———. "Cosmopolitans and Celebrities." *Race & Class* 31.1 (Fall 1989): 1–19.
Buhle, Paul, ed. *C. L. R. James: The Man and His Work.* Special issue of *Urgent Tasks* 12 (1981).
Burton, Antoinette. *At the Heart of Empire: Indians and the Colonial Encounter in Late-Victorian England.* Berkeley: University of California Press, 1996.
Cabral, Amilcar. "National Liberation and Culture." *Return to the Source: Selected Speeches by Amilcar Cabral.* Edited by African Information Service. New York: Monthly Review Press, 1973: 30–56.
———. "The Weapon of Theory." *Revolution in Guinea: Selected Texts by Amilcar Cabral.* Translated and edited by Richard Handyside. New York: Monthly Review Press, 1969: 90–111.
Cartelli, Thomas. "After *The Tempest:* Shakespeare, Postcoloniality, and Michelle Cliff's New, New World Miranda." *Contemporary Literature* 36. 1 (1995): 82–102.
Chatterjee, Partha. *The Nation and its Fragments: Colonial and Postcolonial Histories.* Princeton: Princeton University Press, 1993.
———. *Nationalist Thought and the Colonial World: A Derivative Discourse.* Minneapolis: University of Minnesota Press, 1993.
Cliff, Michelle. *No Telephone to Heaven.* New York: Plume, Penguin Group, 1996. (Originally published in 1987)
———. *Abeng.* New York: Penguin Books, 1991. (Originally published in 1984)
———. *Free Enterprise.* New York: Dutton, 1993.
———. "Journey into Speech—A Writer Between Two Worlds: An Interview with Michelle Cliff." By Opal Palmer Adisa. *African American Review* 28. 2 (1994): 273–281.
———. "An Interview with Michelle Cliff." By Meryl Schwartz. *Contemporary Literature* 34. 4 (1993): 595–619.

———. "The Art of History: An Interview with Michelle Cliff." By Judith Raiskin. *Kenyon Review* 15. 1 (1993): 57–71.

———. "Clare Savage as Crossroads Character." *Caribbean Women Writers: Essays from the First International Conference.* Ed. Selwyn R. Cudjoe. Wellesley, Mass.: Calaloux Publications, 1990.

———. *The Land of Look Behind: Prose and Poetry.* Ithaca, N.Y.: Firebrand Books, 1985.

Clifford, James. "Diasporas." *Cultural Anthropology* 9.3 (1994): 302–338.

———. "The Politics of Neologism: Aimé Césaire." *The Predicament of Culture: Twentieth-Century Ethnography, Literature, and Art.* Cambridge, Mass.: Harvard University Press, 1988: 175–181.

Cohen, Robin. *Global Diasporas: An Introduction.* Seattle: University of Washington Press, 1997.

Colley, Linda. "Britishness and Otherness: An Argument." *Journal of British Studies* 31.4 (1992): 309–329.

Collier, Eugenia. "C. L. R. James." *Fifty Caribbean Writers: A Bio-Bibliographical Critical Source Book.* Ed. Daryl Cumber Dance. Westport, Conn.: Greenwood Press, 1986: 229–238.

Cooper, Brenda. "*Chaiba the Algerian* versus *Our Sister Killjoy:* The Case for Materialist Black Aesthetic." *English in Africa* 12. 2 (1985): 21–51.

Cooper, Frederick. "Conflict and Connection: Rethinking Colonial History." *American Historical Review* (December 1994): 1516–1545.

Crick, Bernard, ed. *National Identities: The Constitution of the United Kingdom.* Oxford: Blackwell Publishers, 1991.

———. "The English and the British." *National Identities:* 90–104.

Dadzie, Stella. "Searching for the Invisible Woman: Slavery and Resistance in Jamaica." *Race & Class* 32 (1990): 21–38.

Dayan, Joan. "Haiti, History, and the Gods." *After Colonialism: Imperial Histories and Postcolonial Displacements.* Ed. Gyan Prakash. Princeton: Princeton University Press, 1995: 66–97.

Dhillon-Kashyap, Perminder. "Locating the Asian Experience." *Screen* 29.4 (Autumn 1985): 120–126.

Diawara, Manthia. "Englishness and Blackness: Cricket as Discourse on Colonialism." *Callaloo* 13 (1990): 830–844.

Dingwaney, Anuradha. "Author(iz)ing *Midnight's Children* and *Shame:* Salman Rushdie's Constructions of Authority." *Reworlding: The Literature of the Indian Diaspora.* Ed. Emmanuel Nelson. Westport, Conn.: Greenwood Press, 1992: 157–168.

Dollimore, Jonathan. *Sexual Dissidence: Augustine to Wilde, Freud to Foucault.* Oxford: Clarendon Press, 1991.

Eagleton, Terry, Frederic Jameson, and Edward W. Said. *Nationalism, Colonialism, and Literature.* Introduction by Seamus Deane. A Field Day Company Book. Minneapolis: University of Minnesota Press, 1990.

Eagleton, Terry. "Nationalism: Irony and Commitment." *Nationalism, Colonialism, Literature:* 23–39.

Ebert, Teresa. "The 'Difference' of Postmodern Feminism." *College English* 53.8 (December 1991): 886–904.

Edmondson, Belinda. *Making Men: Gender, Literary Authority, and Women's Writing in Caribbean Narrative.* Durham: Duke University Press, 1999.

———. "Race, Privilege, and the Politics of (Re)Writing History: An Analysis of the Novels of Michelle Cliff." *Callalloo* 16. 1 (1993): 180–191.

Fanon, Frantz. *Black Skin, White Masks.* Translated from the French by Charles Lam Markmann. New York: Grove Press, Inc., 1967.

———. "The Pitfalls of National Consciousness." In *The Wretched of the Earth.* Translated from the French by Constance Farrington. New York: Grove Press, Inc., 1966.

Farred, Grant, ed. *Rethinking C. L. R. James.* Oxford: Blackwell Publishers, 1996.

Fisher, Jean. "Editorial: Some Thoughts on 'Contaminations.'" *Third Text* 32 (Autumn 1995): 3–8.

Foucault, Michel. *The History of Sexuality. Vol. 1: An Introduction.* New York: Random House, 1978.

Frankenburg, Ruth, and Lata Mani. "Crosscurrents, Crosstalk: Race, 'Post-coloniality,' and the Politics of Location." *Cultural Studies* 7 (May 1993): 292–309.

Gandhi, M. K. *Hind Swaraj or Indian Home Rule.* Ahmedabad: Navjivan Publishing House, 1938.

Geiger, Susan. "Women and African Nationalism." *Journal of Women's History* 2.1 (Spring 1990): 227–244.

Gikandi, Simon. *Maps of Englishness: Writing Identity in the Culture of Colonialism.* New York: Columbia University Press, 1996.

Gilbert, Harriet. "Concrete Jungle Book." *Listener,* vol. 123 (April 1990): 2.

Gilroy, Paul. *There Ain't No Black in the Union Jack: The Cultural Politics of Race and Nation.* London: Hutchinson, 1987.

Grewal, Inderpal. "Salman Rushdie: Women, Marginality, and *Shame.*" *Genders* 3 (1990): 24–42.

Grewal, Inderpal, and Caren Kaplan. *Scattered Hegemonies: Postmodernity and Transnational Feminist Practices.* Durham: Duke University Press, 1994.

Guha, Ramchandra. "Cricket and Black Nationalism." *The Sunday Observer* (New Delhi, India), May 27, 1990: 6.

Guha, Ranajit. "The Small Voice of History." *Subaltern Studies* IX (1996): 1–12.

Hall, Stuart. *Stuart Hall: Critical Dialogues in Cultural Studies.* Ed. David Morley and Kuan-Hsing Chen. London and New York: Routledge, 1996.

———. "The Problem of Ideology: Marxism Without Guarantees." *Stuart Hall:* 25–46.

———. "For Allon White." *Stuart Hall:* 287–305.

———. "New Ethnicities." *Stuart Hall:* 441–449.

———. "The Formation of the Diasporic Intellectual: An Interview with Stuart Hall by Kuan-Hsing Chen." *Stuart Hall:* 484–503.

———. "The After-life of Frantz Fanon: Why Fanon? Why Now? Why *Black Skin, White Masks?*" *The Fact of Blackness: Frantz Fanon and Visual Representation.* London: Institute of Contemporary Arts, 1996: 12–37.

———. "Minimal Selves." Houston A. Baker Jr.: 114–119.

Harding, Sandra. "Subjectivity, Experience, and Knowledge: An Epistemology from/for Rainbow Coalition Politics." *Who Can Speak? Authority and Critical Identity.* Ed. Judith Roof and Robyn Wiegman. Urbana and Chicago: University of Illinois Press, 1995: 120–136.

Hashmi, Alamgir. "Hanif Kureishi and the Tradition of the Novel." *Critical Survey* 5.1 (1993): 25–33.

Haynes, Douglas, and Gyan Prakash, eds. *Contesting Power: Resistance and Everyday Social Relations.* Berkeley and Los Angeles: University of California Press, 1992.

Hebdige, Dick. "Digging for Britain: an Excavation in Seven Parts." *Come on Down: Popular Media Culture in Postwar Britain.* Ed. Dominic Strinatti and Stephen Wagg. London: Routledge, 1992.

Hill, Robert. "In England, 1932–1938." Buhle: 19–27.

Hobsbawm, Eric, and Terence Ranger. *The Invention of Tradition.* Cambridge: Cambridge University Press, 1983.

hooks, bell. "Narratives of Struggle." *Critical Fictions: The Politics of Imaginative Writing.* Ed. Philomena Mariani. Seattle: Bay Press, 1991: 53–61.

———. "Stylish Nihilism: Race, Sex, and Class at the Movies." *Yearning: Race, Gender, and Cultural Politics.* Boston: Southend Press, 1990: 155–163.

Innes, C. L. "Mothers or Sisters? Identity, Discourse and Audience in the Writing of Ama Ata Aidoo and Mariama Bâ." *Motherlands: Black Women's Writing from Africa, the Caribbean, and South Asia.* Ed. Sushila Nasta. London: The Women's Press: 129–151.

Irele, Abiola. "In Praise of Alienation." Mudimbe: 201–224.

James, Adeola, ed. "Ama Ata Aidoo." *In Their Own Voices: African Women Writers Talk.* London: James Currey, 1990: 9–27.

James, C. L. R. *Spheres of Existence: Selected Writings.* Westport, Conn.: Lawrence Hill & Co., 1980.

———. "Two Young Americans." *Spheres:* 106–112.

———. "The Making of the Caribbean People." *Spheres:* 173–190.

———. "Black Power." *Spheres:* 221–236.

———. "Discovering Literature in Trinidad." *Spheres:* 237–244.

———. *Beyond a Boundary.* London: Stanley Paul, 1963.

———. "Foreword." *Beyond a Boundary.* London: Allison and Busby, 1980: v–viii.

———. *The Black Jacobins: Toussaint L'Ouverture and the San Domingo Revolution.* 2nd ed. New York: Vintage Books, 1963.

———. *Kas-Kas: Interviews with Three Caribbean Writers in Texas: George Lamming, C. L. R. James, and Wilson Harris.* Ed. Ian Munro and Reinhard Sander. Austin: The University of Texas African and African American Research Institute, 1972: 23–40.

James, Winston. "The Making of Black Identities." Samuel, vol. 2: 230–255.

Jameson, Fredric. "Modernism and Imperialism." *Nationalism, Colonialism, Literature:* 44–65.

JanMohamed, Abdul R. *Manichean Aesthetics: The Politics of Literature in Colonial Africa.* Amherst: University of Amherst Press, 1983.

———. "Wordliness-Without-World, Homelessness-Without-Home: Toward a Definition of a Specular Border Intellectual." Sprinker: 96–120.

———. "The Economy of Manichean Allegory: The Function of Racial Difference in Colonialist Literature." *Critical Inquiry* 12 (1985): 59–87.

Jayawardena, Kumari. *Feminism and Nationalism in the Third World.* New Delhi: Kali for Women, 1986.

Jewsiewicki, Bogumil. "*Présence Africaine* as Historiography: Historicities of Society and Specificity of Black African Culture." Mudimbe: 118–135.

Jordan, June. *Civil Wars.* Boston: Beacon Press, 1981.

Joshi, Svati, ed. *Rethinking English: Essays on Literature, Language, History.* New Delhi: Trianka, 1991.

Jussawalla, Feroza. *Family Quarrels: Towards a Criticism of Indian Writing in English.* New York: Peter Lang, 1985.

———. "Resurrecting the Prophet: The Case of Salman, the Otherwise." *Public Culture* 2 (Fall 1989): 106–117.

Kaleta, Kenneth C. *Hanif Kureishi: Postcolonial Storyteller.* Austin: University of Texas Press, 1998.

Kaplan, Caren. "Deterritorializations: The Rewriting of Home and Exile in Western Feminist Discourse." *Cultural Critique* 6 (Spring 1997): 188–196.

Karamcheti, Indira. "Aimé Césaire's Subjective Geographies: Translating Place and the Difference it Makes." *Between Languages and Cultures: Translation and Cross-Cultural Texts.* Ed. Anuradha Dingwaney and Carol Maier. Pittsburgh: University of Pittsburgh Press, 1995: 181–197.

Katrak, Ketu. "Afterword," *No Sweetness Here and Other Stories* by Ama Ata Aidoo. New York: The Feminist Press, 1995: 135–160.

Kearney, Hugh. "Four Nations or One?" *National Identities:* 1–6.

Krishnaswamy, Revathi. "Mythologies of Migrancy: Postcolonialism, Postmodernism and the Politics of (Dis)location. *Ariel: A Review of International English Literature* 26.1 (1995): 125–146.

Kureishi, Hanif. *The Buddha of Suburbia.* Penguin Books, 1991.

————. *London Kills Me: Three Screenplays and Four Essays.* Penguin Books, 1992.

————. *The Black Album.* Scribner Paperback Edition, 1996.

————. "The Rainbow Sign." *London Kills Me:* 1–37.

————. "Some Time with Stephen: A Diary." *London Kills Me:* 117–194.

LaCapra, Dominick, ed. *The Bounds of Race: Perspectives on Hegemony and Resistance.* Ithaca: Cornell University Press, 1991.

Lazarus, Neil. *Resistance in Postcolonial African Fiction.* New Haven and London: Yale University Press, 1990.

———. "National Consciousness and the Specificity of (Post-)Colonial Intellectualism." Francis Barker et al: 197–220.

Lima, Helen Maria. "Revolutionary Developments: Michelle Cliff's 'No Telephone to Heaven' and Merle Collins's 'Angel.'" *Ariel: A Review of International English Literature* 24.1 (1993): 35–56.

Lionnet, Françoise. "Of Mangoes and Maroons: Language, History, and the Multicultural Subject of Cliff's *Abeng.*" *De/Colonizing the Subject: The Politics of Gen-*

der in Women's Autobiography. Ed. Sidonie Smith and Julia Watson. Minneapolis: University of Minnesota Press, 1992: 321–345.

Lipscomb, David. "Caught in a Strange Middle Ground: Contesting History in Salman Rushdie's *Midnight's Children.*" *Diaspora: A Journal of Transnational Studies* 1 (1991): 163–189.

Loomba, Ania. "Overworlding the 'Third World'." *Oxford Literary Review* 13.1–2 (1991): 164–91

Loomba, Ania, and Suvir Kaul. "Introduction: Location, Culture, Post-coloniality." *On India: Writing History, Culture, Post-Coloniality.* Ed. Ania Loomba and Suvir Kaul. Special Issue of *Oxford Literary Review* 16 (1994): 3–30.

MacKinnon, Catherine A. "Feminism, Marxism, Method, and the State: An Agenda for Theory." *Signs: Journal of Women in Culture and Society* 7. 3 (1982): 515–544.

Maharaj, Sarat. "The Congo is Flooding the Acropolis: Black Art, Orders of Difference, Textiles." *Interrogating Identity.* New York: Grey Art Gallery and Study Center, 1991: 13–42.

McClintock, Anne. "The Angel of Progress: Pitfalls of the Term 'Post-Colonialism.'" *Social Text* 31/32 (1992): 84–98.

———. "'No Longer in a Future Heaven': Women and Nationalism in South Africa." *Transition* 51 (1991): 104–123.

Mercer, Kobena. "Recoding the Narratives of of Race and Nation." *The Independent* (January/February, 1989): 19–26.

————. "Diaspora Culture and the Dialogic Imagination." *Black Frames: Critical Perspectives on Black Independent Cinema.* Ed. Mbye Cham and Claire Andrade-Watkins. Cambridge, Mass: M.I.T. Press, 1988: 50–61.

Mill, John Stuart. "Coleridge." *The Norton Anthology of Literature.* 6th ed. Vol. 2: 1001–1003.

Miller, Christopher L. *Nationalists and Nomads: Essays on Francophone Literature and Culture.* Chicago: University of Chicago Press, 1998.

———. "Alioune Diop and the Unfinished Temple of Knowledge." Mudimbe: 427–343.

Miller, Daniel, Michael Rowlands, and Christopher Tilley. *Domination and Resistance.* London and Boston: Allen and Unwin, 1988.

Modebe, Sarah. "Ama Ata Aidoo—In Conversation." *New African* 288 (Sept. 1991): 40.

Modood, Tariq. "'Difference,' Cultural Racism and Anti-Racism." Werbner and Modood: 154–172.

Mohanram, Radhika. "Postcolonial Spaces and Deterritorialized (Homo) Sexuality: The Films of Hanif Kurieshi." *Postcolonial Discourse and Changing Cultural Contexts: Theory and Criticism.* Ed. Gita Rajan and Radhika Mohanram. Westport, Conn.: Greenwood Press, 1995: 117–134.

Mohanty, Chandra Talpade. "Feminist Encounters: Locating the Politics of Experience." *Copyright* 1 (Fall 1987): 30–44.

Mouralis, Bernard. "*Présence Africaine:* Geography of an Ideology." Mudimbe: 3–13.

Morley, David and Kevin Robins. "Spaces of Identity: Communications Technologies and the Reconfiguration of Europe." *Screen* 30.4 (Autumn 1989): 10–34.

Morris, Meaghan. "Brasschaat-Bombay: A Way of Inhabiting a Culture." Introduction to Paul Willemen's *Looks and Friction:* 1–23.

Mudimbe, V.Y. "Finale." Mudimbe: 435–445.

Mudimbe, V.Y. ed. *The Surreptitious Speech: Présence Africaine and the Politics of Otherness, 1947–1987.* Chicago: University of Chicago Press, 1992.

Nandy, Ashis. *The Tao of Cricket: On Games of Destiny and the Destiny of Games.* New Delhi: Penguin Books, 1989.

———. "Shamans, Savages, and the Wilderness: On the Audibility of Dissent and the Future of Civilizations." *Alternatives* 14 (1989): 263–77.

Needham, Anuradha Dingwaney. "Fictions of National Belonging." *Socialist Review* 24.4 (1994): 173–183.

———. "Reimagining Familiar Dichotomies in Reading 'Alternative' Texts." *The Journal of the Midwest Modern Modern Language Association* 25.1 (Spring 1992): 47–53.

Nehru, Jawaharlal. *The Discovery of India.* Delhi: Oxford University Press, 1994.

Odamtten, Vincent O. *The Art of Ama Ata Aidoo: Polylectics and Reading Against Neo-Colonialism.* Gainesville: University Press of Florida, 1994.

O'Hanlon, Rosalind. "Cultures of Rule, Communities of Resistance: Gender, Discourse, and Tradition in Recent South Asian Historiographies." *Social Analysis* 25 (1989): 94–114.

———. "Recovering the Subject: *Subaltern Studies* and Histories of Resistance in South Asia." *Modern Asian Studies* 21 (1988): 189–224.

Okihiro, Gary, ed. *In Resistance: Studies in African, Caribbean, and Afro-American History.* Amherst: University of Massachusetts Press, 1986.

Oldenburg, Veena Talwar. "Lifestyle as Resistance: The Case of the Courtesans of Lucknow, India." *Feminist Studies* 16 (1990): 259–287.

Ortner, Sherry. "Resistance and the Problem of Ethnographic Refusal." *The Historic Turn in Human Sciences.* Ed. Terence J. MacDonald. Ann Arbor: University of Michigan Press, 1996: 281–304.

Pally, Marcia. "Kureishi Like a Fox." *Film Comment* 22.5 (Sept./Oct. 1986): 50–55

Pandey, Gyanendra. *The Construction of Communalism in Colonial North India.* Delhi: Oxford University Press, 1992.

Papastergiadis, Nikos. "Restless Hybrids." *Third Text* 32 (Autumn 1995): 9–18.

Parry, Benita. "Signs of Our Times: Discussion of Homi Bhabha's *The Location of Culture.*" *Third Text* 28/29 (Autumn/Winter 1994): 5–24.

———. "Resistance Theory/Theorizing Resistance or Two Cheers for Nativism." Francis Barker et al.: 172–196.

———. "Overlapping Territories and Intertwined Histories: Edward Said's Postcolonial Cosmopolitanism." Sprinker: 19–47.

———. "Problems in Current Theories of Colonial Discourse." *Oxford Literary Review* 9. 1–2 (1987): 27–58.

Pérez Firmat, Gustavo. *The Cuban Condition: Translation and Identity in Modern Cuban Literature.* Cambridge: Cambridge University Press, 1989.

Phillips, Caryl, ed. *Extravagant Strangers: A Literature of Belonging.* London and Boston: Faber and Faber, 1997.

Prakash, Gyan. "Postcolonial Criticism and Indian Historiography." *Social Text* 31/32 (1992): 8–19.

———. "Writing Post-Orientalist Histories of the Third World: Perspectives From Indian Historiography." *Comparative Studies in Society and History* 32 (1990): 383–408.

Pratt, Mary Louise. *Imperial Eyes: Travel Writing and Transculturation.* London and New York: Routledge, 1992.

———. "The Arts of the Contact Zone." *Profession 91.* The Modern Language Association of America: 33–40.

Ramos, Alcida. "Indian Voices: Contact Experienced and Expressed." *Rethinking History and Myth: Indigenous South American Perspectives on the Past.* Ed. Jonathan Hill. Chicago: University of Illinois Press, 1988: 214–234.

Robinson, Cedric J. *Black Marxism.* London: Zed Press, 1983.

———. "Amilcar Cabral and the Dialectic of Portuguese Colonialism." *Radical America* (May-June 1981): 39–57.

Rodney, Walter. "The African Revolution." Buhle: 5–12.

Rushdie, Salman. *The Moor's Last Sigh.* New York: Pantheon Books, 1995.

———. *The Satanic Verses.* New York: Viking Penguin, Inc., 1989.

———. *Shame.* New York: Vintage/Aventura, 1984.

———. *Midnight's Children.* New York: Avon Books, 1980.

———. *"Midnight's Children and Shame."* Kunapipi 7 (1985): 1–19.

———. "The Location of Brazil." *American Film* 10 (1985): 50–53.

———. Interview with David Brooks. *Helix* 19 & 20 (1985): 55–69.

———. "A Dangerous Art Form." Interview with *Third World Book Review* 1 (1984): 3–5.

———. "The Indian Writer in England." *The Eye of the Beholder: Indian Writing in English.* Ed. Maggie Butcher. London: Commonwealth Institute, 1983: 75–83.

———. "Outside the Whale." *Granta* 11 (1983): 123–141.

———. "Interview with Rani Dharker." *New Quest* 42 (1983): 351–360.

———. "The New Empire within Britain." *New Society* (Dec. 9, 1982): 34–37.

Safran, William. "Diasporas in Modern Societies: Myths of the Homeland and Return." *Diaspora: A Journal of Transnational Studies* 1.1 (1991): 83–99.

Said, Edward W. *Culture and Imperialism.* New York: Alfred A. Knopf, 1993.

———. *Orientalism.* New York: Vintage Books, 1979.

———. "Third World Intellectuals and Metropolitan Culture." *Raritan* 9 (1990): 27–50.

———. "Yeats and Decolonization." *Remaking History.* Eds. Barbara Kruger and Phil Mariani. Seattle: Bay, 1989: 3–29.p

———. "Intellectuals in the Postcolonial World." *Salmagundi* 70/71 (1986): 44–64.

———. "In the Shadow of the West." *Wedge* 7/8 (Winter-Spring 1985): 4–11.

———. "Travelling Theories." *Raritan* 1 (1982): 41–67.

Samuel, Raphael, ed. *Patriotism: The Making and Unmaking of British National Identity. Vol I: History and Politics.* London and New York: Routledge, 1989.

———. ed. *Patriotism: The Making and Unmaking of British National Identity Vol II: Minorities and Outsiders.* London and New York: Routledge, 1989.

———. "Preface." *Patriotism Vol I:* x-xvii

———. "Introduction: Exciting to be English." *Patriotism Vol. I:* xviii-lxiii.

Samantrai, Ranu. "Caught at the Confluence of History: Ama Ata Aidoo's Necessary Nationalism." *Research in African Literatures* 26.2 (Summer 1995): 140–157.

Sander, Reinhard. *Trinidad Awakening: West Indian Literature in the Nineteen Thirties.* New York: Greenwood Press, 1988.

Sangari, KumKum. "The Politics of the Posssible." *The Nature and Context of Minority Discourse.* Ed. Abdul JanMohamed and David Lloyd. New York: Oxford University Press, 1990: 216–245.

Sangari, KumKum, and Sudesh Vaid. *Recasting Women: Essays in Colonial History.* New Delhi: Kali for Women, 1989.

Scott, James C. *Domination and the Arts of Resistance: Hidden Transcripts.* New Haven and London: Yale University Press, 1990.

Scott, Joan W. "Gender: A Useful Category of Historical Analysis." *The American Historical Review* 91 (1986): 1053–1075.

Searle, Chris. *Words Unchained: Language and Revolution in Grenada.* London: Zed Books, 1984.

———. "Race before Wicket: Cricket, Empire, and the White Rose." *Race & Class* 31 (1990): 31–48.

Sharpe, Jenny. "Figures of Colonial Resistance." *Modern Fiction Studies* 35.1 (Spring 1989): 137–155.

Selvon, Sam. *The Lonely Londoners.* Essex: Longman Group 1956.

Sinfield, Alan. *Faultlines: Cultural Materialism and the Politics of Dissident Reading.* Berkeley: University of California Press, 1992.

Sivanadan, A. "The Enigma of the Colonised: Reflections on Naipaul's Arrival." *Race & Class* 32.1 (1990): 33–43.

Smith, Faith L. "Coming Home to the Real Thing: Gender and Intellectual Life in the Anglophone Caribbean." *The South Atlantic Quarterly* 93.4 (1994): 895–923.

Spivak, Gayatri Chakravorty. *Outside in the Teaching Machine.* New York and London: Routledge, 1993.

———. (with Ellen Rooney). "In a Word: Interview." *Outside:* 1–24.

———. "Marginality in the Teaching Machine." *Outside:* 53–76.

———. "Feminism and Deconstruction, Again: Negotiations." *Outside:* 121–140.

———. "Sammy and Rosie Get Laid." *Outside:* 243–254.

———. *The Post-Colonial Critic: Interviews, Strategies, Dialogues.* Ed. Sarah Harasym. New York and London: Routledge, 1990.

———. (with Elizabeth Grosz). "Criticism, Feminism, and the Institution." *The Post-Colonial Critic:* 1–16.

———. *In Other Worlds: Essays in Cultural Politics.* New York: Routledge, 1988.

———. "Translator's Foreword" to "Draupadi" by Mahasweta Devi and "A Literary Representation of the Subaltern: A Woman's Text from the Third World." *In Other Worlds:* 179–187 and 241–268.

———. "The Burden of English." *The Lie of the Land: English Literary Studies in India.* Ed. Rajewari SunderRajan. Delhi: Oxford University Press, 1992: 275–299.

———. "Teaching for the Times." *The Journal of the Midwest Modern Language Association* 25.1 (Spring 1992): 3–22.

———. "Theory in the Margins: Coetzee's Foe Reading Defoe's Crusoe/Roxana." *Consequences of Theory.* Ed. Jonathan Arac and Barbara Johnson. Baltimore: Johns Hopkins University Press, 1991: 154–181.

———. "Who Claims Alterity?" *Remaking History.* Ed. Barbara Kruger and Phil Mariani. Seattle: Bay Press, 1989: 269–292.

———. "Discussion: Subaltern Studies: Deconstructing Historiography." *Subaltern Studies* IV (1985): 330–363.

Sprinker, Michael, ed. *Edward Said: A Reader.* Oxford and Cambridge, Mass.: Blackwell, 1992.

Stepan, Nancy Leys and Sander Gillman. "Appropriating the Idioms of Science: The Rejection of Scientific Racism." LaCapra: 72–103.

Stratton, Jon and Ien Ang. "On the Impossibility of a Global Cultural Studies: British Cultural Studies in an 'International Frame.'" *Stuart Hall:* 361–391.

Suleri, Sara. *The Rhetoric of English India.* Chicago: University of Chicago Press, 1994.

Tawadros, Gilane. "Other Britains, Other Britons." *Aperture* (San Franscisco) 113 (1988): 40–46.

Terdiman, Richard. *Discourse/Counter-Discourse: The Theory and Practice of Symbolic Resistance in Nineteenth-Century France.* Ithaca: Cornell University Press, 1985.

Tharu, Susie, and K. Lalitha, eds. *Women Writing in India, 600 B.C. to the Present.* 2 vols. New York: The Feminist Press, 1991.

———. "The Arrangement of an Alliance: English Studies and the Making of Indian Literatures." Joshi: 160–180.

Tölölyan. Khachig. "The Nation-State and its Others: In Lieu of a Preface." *Diaspora: A Journal of Transnational Studies* 1.1 (1991): 3–7.

Thomas, Richard. "Black Scholar." Buhle: 121.

Vaillant, Janet G. *Black, French, and African: A Life of Léopold Sidar Senghor.* Cambridge, Mass.: Harvard University Press, 1990.

Viswanathan, Gauri. *Masks of Conquest: Literary Study and English Rule in India.* New York: Columbia University Press, 1989.

———. "Raymond Williams and British Colonialism." *The Yale Journal of Criticism* 4.2 (Spring 1991): 47–66.

Visweswaran, Kamala. *Fictions of Feminist Ethnography.* Minneapolis: University of Minnesota Press, 1994.

Walters, Margaret. "Kureishi's England." *The London Review of Books,* 5 April, 1990: 16–17.

White, E. Frances. "Africa on My Mind: Gender, Counter Discourse and African-American Nationalism." *Journal of Women's History* 2.1 (Spring 1990): 73–97.

White, Hayden. *Tropics of Discourse.* Baltimore: Johns Hopkins University Press, 1978.

Willemen, Paul. *Looks and Frictions: Essays in Cultural Studies and Film Theory.* Bloomington and Indianapolis: Indiana University Press, 1994.

Williams, Raymond. *Resources of Hope: Culture, Democracy, Socialism* Ed. Robin Gale. Introduction by Robin Blackburn. London and New York: Verso, 1988.

———. *Marxism and Literature.* Oxford: Oxford University Press, 1977.

———. "Media, Margins, and Modernity: A Conversation between Raymond Williams and Edward Said," Appendix to Raymond Williams, *The Politics of Modernism: Against the New Conformists.* Edited and Introduced by Tony Pinkney. London: Verso, 1989: 177–198.

Wynter, Sylvia. "The Ceremony Must Be Found: After Humanism." *boundary 2* (1984): 19–70.

———. "In Quest of Matthew Bondsman: Some Cultural Notes on the Jamesian Journey." Buhle: 54–68.

Yousaf, Nahem. "Hanif Kureishi and 'the Brown Man's Burden.'" *Critical Survey* 8.1 (1996): 14–25.

Index